Modeling Fluctuations in Scattered Waves

Series in Optics and Optoelectronics

Series Editors: **R G W Brown**, University of Nottingham, UK
E R Pike, Kings College, London, UK

Other titles in the series

Series in Optics and Optoelectronics

Modeling Fluctuations in Scattered Waves

E Jakeman
University of Nottingham, UK

K D Ridley
QinetiQ, Worcestershire, UK

Taylor & Francis
Taylor & Francis Group
New York   London

Taylor & Francis is an imprint of the
Taylor & Francis Group, an informa business

CRC Press
Taylor & Francis Group
6000 Broken Sound Parkway NW, Suite 300
Boca Raton, FL 33487-2742

© 2006 by Taylor and Francis Group, LLC
CRC Press is an imprint of Taylor & Francis Group, an Informa business

No claim to original U.S. Government works
Printed in the United States of America on acid-free paper
10 9 8 7 6 5 4 3 2 1

International Standard Book Number-10: 0-7503-1005-7 (Hardcover)
International Standard Book Number-13: 978-0-7503-1005-5 (Hardcover)

Visit the Taylor & Francis Web site at
http://www.taylorandfrancis.com

and the CRC Press Web site at
http://www.crcpress.com

Preface

Fluctuations in scattered light constitute some of the most familiar natural optical phenomena, including the twinkling of starlight, the glittering of sunlight on a rippled water surface, and the shimmering of distant objects on a hot day. These effects are geometrical in origin, being a consequence of refraction by large-scale random variations in refractive index. Fluctuations of this kind may be caused by density variations within a propagation medium, such as the atmosphere, or by roughness of the interface between media having differing dielectric constants. Although significant dispersion may also be observed (for example, in the case of the colorful twinkling of a star low in the sky), for the most part they are white light optical effects that are independent of wavelength within the visible spectrum. With the advent of the laser, a new range of coherent-light scattering phenomena became visible. In particular, the random interference effect that has come to be known as *laser speckle* commonly occurs in laboratory laser light scattering experiments (Figure 0.1). Indeed, laser light scattering generates many beautiful diffraction, interference, and geometrical optics effects that can be observed with the naked eye. Figure 0.2 shows the effect of scattering multifrequency laser light from a piece of ground glass in the laboratory. Figure 0.3 shows the optical intensity pattern obtained when laser light is passed through turbulent air rising above a heating element, and is rather reminiscent of the pattern sometimes observed on the floor of a swimming pool.

It is important to recognize that a similar range of effects is generated at other frequencies of the electromagnetic spectrum and in the scattering of sound waves. These can and have been measured, but of course they cannot be seen with the naked eye. However, they obviously contain information about the scattering object and they also limit the performance of sensing and communication systems. One familiar example of this is the fading of shortwave radio reception due to ionospheric fluctuations. Many early theoretical results were derived as a consequence of the observation of fluctuations at radio and radar wavelengths. For example, an early theoretical description of random interference effects was developed to explain the fluctuating radar return from raindrops twenty years before the visual appearance of the equivalent laser-generated phenomenon gave rise to the term "speckle." It was soon recognized that intensity patterns of the form shown in Figure 0.1 were manifestations of Gaussian noise, a well-known statistical model that is readily characterized and amenable to calculation. Unfortunately, in practice the more complicated radiation patterns illustrated in Figure 0.2 and Figure 0.3 are common and the Gaussian noise model

FIGURE 0.1
Random interference pattern generated when laser light is scattered by ground glass. (See color insert following page 14.)

cannot adequately describe these. The aim of this book is to provide a practical guide to the phenomenology, mathematics, and simulation of non-Gaussian noise models and how they may be used to characterize the statistics of scattered waves.

The plan of the book is as follows: After Chapter 1, in which the statistical tools and formalism are established, Chapter 2 reviews the properties of Gaussian noise including some lesser known results on the phase statistics. Chapter 3 describes processes derived from Gaussian Noise that are commonly encountered, while Chapter 4 discusses deviation from Gaussian statistics in the context of the random walk model for discrete scattering centers. The random phase-changing screen is a scattering system that introduces random distortions into an incident wave front. It provides an

FIGURE 0.2
The effect of focusing multifrequency laser light onto a small area of the same kind of scatterer as in Figure 0.1. (See color insert following page 14.)

excellent model for a wide variety of continuum scattering systems, ranging from ionospheric scattering of radio waves to light scattering from rough surfaces, and is a valuable aid to understanding the phenomenology in non-Gaussian scattering regimes. Chapter 5 through Chapter 8 provides an overview of the predictions of this ubiquitous model, making various assumptions for the properties of the initial phase distortion. Chapter 9 addresses aspects of propagation through an extended medium while Chapter 10 discusses some multiple scattering effects. The scattering of vector waves and polarization fluctuations are discussed in Chapter 11. Chapter 12 is devoted to a discussion of perhaps the most widely used non-Gaussian model: K-distributed noise. Chapter 13 outlines some of the practical limitations encountered in experimental measurement and how they affect the interpretation of results, detection, and measurement accuracy. Finally, Chapter 14 will describe techniques for numerical simulation.

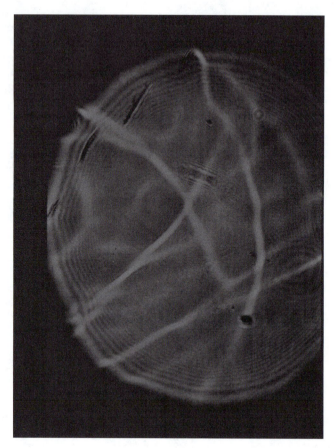

FIGURE 0.3
Intensity pattern obtained when laser light is passed through air convecting above a heating element. (See color insert following page 14.)

Acknowledgments

The authors would like to acknowledge the invaluable contributions of their coworkers to the topics covered by this book, particularly those of their immediate colleagues and former colleagues at Malvern and Nottingham: Roy Pike, Peter Pusey, John McWhirter, John Jefferson, Gareth Parry, John Walker, Dave Jordan, Robert Tough, Keith Ward, Paul Tapster, Chris Hill, and Keith Hopcraft.

Authors

Eric Jakeman is an emeritus professor of the University of Nottingham, associated with the School of Electrical and Electronic Engineering and the School of Mathematical Sciences. He graduated with a degree in mathematical physics at Birmingham University in 1960 and obtained his PhD on superconductivity theory in 1963. After a year researching low-energy nuclear physics at the University of California, Los Angeles, he joined the Royal Radar Establishment (now QinetiQ) Malvern in 1964 where he eventually rose to the position of chief scientist in the electronics business sector. His interests have ranged from heat and mass transfer problems in crystal growth to photon-counting statistics, but since the mid-1970s his principal area of work has been the analysis of fluctuations in scattered waves. Professor Jakeman has published more than 180 papers in refereed journals. He was elected Fellow of the Institute of Physics in 1979, Fellow of the Optical Society of America in 1988, and Fellow of the Royal Society of London in 1990.

Professor Jakeman was a member of the Council of the Institute of Physics from 1985 until 2003 and was vice president for publications and chairman of the publishing board 1989 to 1993 and again 2002 to 2003. He was honorary secretary from 1994 to 2003. He has also served on the executive committee of the European Physical Society. University positions held include visiting professor in the physics department at Imperial College of Science Technology and Medicine, London, and special professor at the University of Nottingham where he took up a permanent position in 1996.

Kevin Ridley is currently employed as a laser physicist at QinetiQ Malvern. He graduated from the University of Bath with a degree in applied physics in 1987. He then joined the Royal Signals and Radar Establishment (now QinetiQ) Malvern and obtained his PhD as an external student at Imperial College London in 1993 on the subject of optical phase conjugation and stimulated Brillouin scattering. He has worked on a wide range of laser-related topics including nonlinear optics, statistical optics, laser light propagation through the atmosphere, laser vibrometry, and photon-counting detectors. Dr. Ridley has published over 40 papers in refereed journals. He is a QinetiQ Fellow and was elected a Fellow of the Institute of Physics in 2002.

Contents

1

Statistical Preliminaries

1.1 Introduction

The purpose of this chapter is to provide a brief introduction to the statistical quantities and notation that will be used in the remainder of the book. There are many excellent texts on probability theory, noise, and stochastic processes, covering the subjects at various depths and levels of sophistication. The treatment here is aimed at a nonspecialist user community and we shall present a straightforward engineering exposition that covers only the essentials required to understand the principles, significance, and application of the statistical models that are described in the following chapters. A more comprehensive treatment, taking the same general approach, is to be found in the book *An Introduction to the Theory of Random Signals and Noise* by Davenport and Root [1], while for the ultimate encyclopedic treatment the reader is referred to *An Introduction to Statistical Communication Theory* by Middleton [2]. Optical engineers may find the treatment given in the excellent treatise *Statistical Optics* by Goodman [3] more to their taste.

1.2 Random Variables

We start with the description of a simple one-dimensional signal $V(t)$ (Figure 1.1) that is a random function of the time t, that is, a random *process*. Fluctuations in the value of V may be the result of many rapidly changing underlying variables or may be the chaotic outcome of a complicated system of nonlinear differential equations. Although the functional dependence of V on time may be random, the result of a measurement of V can be expressed statistically in terms of the single-time probability density function (pdf) $P(V(t))$ which defines the likelihood of obtaining a value V at time t. Similarly, the double-time density $P(V_1(t_1), V_2(t_2))$ defines the joint probability of obtaining the values V_1 and V_2 by making measurements at times t_1 and t_2,

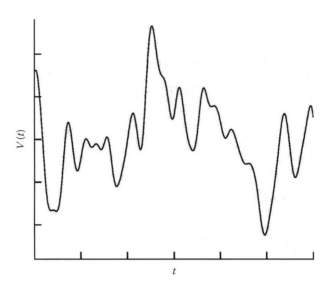

FIGURE 1.1
A random process.

respectively. When two or more random variables are linked in this way, the probability of obtaining a certain value of one irrespective of the value of the other is given by the *marginal* density, which is

$$P(V_1) = \int P(V_1, V_2) dV_2 \qquad (1.1)$$

The probability of obtaining a certain value of one variable given that the other has a known value is called the *conditional* density, and is

$$P(V_1 | V_2) = \frac{P(V_1, V_2)}{P(V_2)} \qquad (1.2)$$

that is, the joint density divided by the marginal density of V_2. Note that we will often use the less formal term *distribution* when referring to a probability density function; this should not be confused with the probability distribution function, which is an integral over the pdf. In order to avoid any confusion we will use the term *cumulative distribution function* when we need to refer to the latter function, which is the probability that the value of a random variable is less than a given value. So

$$F_V(x) = \int_{-\infty}^{x} P(V) dV \qquad (1.3)$$

gives the probability of V having a value less than x.

A knowledge of the entire set of multiple joint distributions $P(V_1(t_1),V_2(t_2),V_3(t_3), \dots) = P(\{V(t_i)\})$ provides a complete statistical description of the quantity $V(t)$. Equivalent information is contained in the moments

$$\langle V^n(t) \rangle = \int_{-\infty}^{\infty} V^n P(V) dV \tag{1.4}$$

and the correlation functions

$$\left\langle \prod_i V^{n_i}(t_i) \right\rangle = \int_{-\infty}^{\infty} P(\{V_i\}) \prod_i V_i^{n_i} dV_i \tag{1.5}$$

The angular brackets denote the integration of the enclosed quantity over one or more probability density functions, the result being referred to as an *average* or *expectation value*.

If the distribution $P(V)$ is "long-tailed," it is possible that some of the quantities defined by Equation 1.4 and Equation 1.5 may not exist, and other measures such as fractional moments will have to be adopted to describe its behavior. However, it is more likely that in practice the finite dynamic range of the measuring apparatus will limit the observation of extreme fluctuations and thereby define a cutoff in the distribution tail.

When the probability distributions do not change with time, so that

$$P\left(\{V(t_i)\}\right) = P\left(\{V(t_i + t)\}\right) \tag{1.6}$$

for all values of t, the process governing the behavior of V is said to be *stationary* and we shall confine our considerations throughout to this kind of process. The property in Equation 1.6 implies, for example, that the autocorrelation function or bilinear moment is a function of the time-difference variable only

$$\langle V(t)V(t+\tau) \rangle = \langle V(0)V(\tau) \rangle \tag{1.7}$$

The formulae in Equation 1.4 and Equation 1.5 constitute formal mathematical definitions but are only useful when a means by which they can be measured in situations of practical interest is identified. In fact, data may be acquired and processed to extract the information contained in the distribution $P(\{V_i\})$, its moments, and its correlation functions in two distinct ways. For example, the moments in Equation 1.4 may be constructed by *time-averaging* powers of V over a single trace. Alternatively, *ensemble averages* may be constructed by averaging values of V taken at a single time from a large number of sample traces. This might be appropriate upon rerunning an experiment several times under the same conditions, for example. When time averages and ensemble averages are the same, so that for a function f

$$\langle f(V) \rangle = \int_{-\infty}^{\infty} f(V)P(V)dV = \lim_{T \to \infty} \frac{1}{T} \int_{-T/2}^{T/2} f(V(t))dt \tag{1.8}$$

the governing process is said to be *ergodic*. Ergodicity requires that all states of the system are accessible to each other through time evolution. Clearly, if this is not the case then some values may not be achieved in a single time trace, although they may be found in a sequence of traces with different initial conditions.

Of course, V may be a function of position rather than time. This could provide a description of the height of a rigid random rough surface that is corrugated in one direction, for example. The representation can be generalized further to include variables that vary in *three* dimensions *and* evolve with time such as the height of a rippling water surface or the random pattern of sunlight reflected from a rippled water surface onto the underside of a bridge. Spatial equivalents of stationarity and ergodicity arise quite naturally in these more general situations.

1.3   Transformation of Variables

Consider again a variable f which is a function of the random variable V. We note that Equation 1.8 may also be written in the form

$$\langle f(V) \rangle = \int_{-\infty}^{\infty} fP(f)df \tag{1.9}$$

using the simple transformation of variables

$$P(f)df = P(V)dV \tag{1.10}$$

Here, $P(V)dV$ can be interpreted as the probability of V having a value between V and $V + dV$. This is the same as the probability of f lying between f and $f + df$; thus

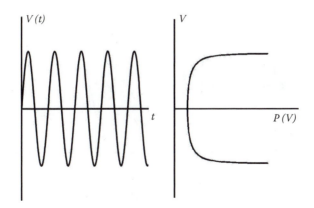

FIGURE 1.2
The pdf of a sinusoidally oscillating variable.

$$P(f) = P(V) \left| \frac{dV}{df} \right| \tag{1.11}$$

Note that the modulus of the derivative is taken so that the pdf is always positive. In fact, Equation 1.11 only applies to the case when f is a single-valued function of V. If N different V values give the same value of f, then a summation of probabilities over those N values is required; that is,

$$P(f) = \sum_{i=1}^{N} P(V_i) \left| \frac{dV_i}{df} \right| \tag{1.12}$$

For example, for a sine wave with phase, ϕ, that is uniformly distributed over the unit circle (Figure 1.2) we have

$$V = A \sin(\omega t + \phi), \qquad P(\phi) = \frac{1}{2\pi} \qquad 0 < \phi < 2\pi$$

Here, V is a dual-valued function of ϕ ; thus the derivative gets multiplied by a factor of two and

$$P(V) = \frac{1}{\pi \sqrt{A^2 - V^2}} \qquad |V| < |A| \tag{1.13}$$

Similarly, if $f = V^2$ and $P(V)$ is a Gaussian distribution (Figure 1.3)

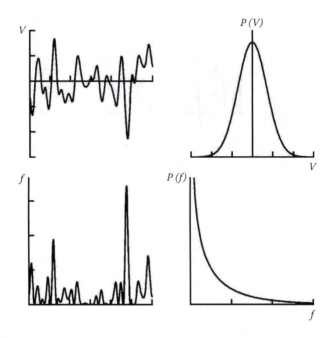

FIGURE 1.3
A Gaussian random process transformed by a square-law device.

$$P(V) = \frac{1}{\sigma\sqrt{2\pi}} \exp\left(-V^2\big/2\sigma^2\right) \qquad -\infty < V < \infty \qquad (1.14)$$

then

$$P(f) = \frac{1}{\sigma\sqrt{2\pi f}} \exp\left(-f\big/2\sigma^2\right) \qquad 0 < f < \infty \qquad (1.15)$$

The output of a square-law device detecting Gaussian noise would have the distribution of Equation 1.15, which is commonly encountered in the radar field [4]. Equation 1.11 can be generalized to effect a transformation between probability distributions of many variables

$$P(\{V_i\})dV_i = |J| P(\{f_i\})df_i \qquad (1.16)$$

where J is the Jacobian for the transformation.

1.4   Wiener-Khinchin Theorem

Consider now the autocorrelation function of a stationary, ergodic process

$$G(\tau) = \langle V(0)V(\tau) \rangle = \lim_{T \to \infty} \frac{1}{T} \int\limits_{-T/2}^{T/2} V(t)V(t+\tau)dt \qquad (1.17)$$

Evidently $G(\tau) = G(-\tau)$. For an entirely random process all memory vanishes for large delay time so that

$$\langle V(0)V(\tau) \rangle \xrightarrow{\ \tau \to \infty\ } \langle V \rangle^2 \qquad (1.18)$$

It is often convenient to define the normalized function

$$g(\tau) = \langle V(0)V(\tau) \rangle \big/ \langle V^2 \rangle \qquad (1.19)$$

Note that the left-hand side of the expression

$$\left\langle \left[V(0) \pm V(\tau) \right]^2 \right\rangle = 2\langle V^2 \rangle \pm 2\langle V(0)V(\tau) \rangle \qquad (1.20)$$

is always greater than or equal to zero, from which it follows that

$$|g(\tau)| \le 1 \qquad (1.21)$$

An arbitrary function $V(t)$ can be expanded in a Fourier series in the interval $[-T/2, T/2]$ provided that the integral

$$\int\limits_{-T/2}^{T/2} V(t)dt \qquad (1.22)$$

exists. We may therefore write

$$V(t) = \sum_{n=-\infty}^{\infty} v_n \exp(-in\omega t) \quad \text{where} \quad \omega = \frac{2\pi}{T}$$

(1.23)

$$\text{and} \quad v_n = \frac{1}{T} \int_{-T/2}^{T/2} V(t) \exp(in\omega t) dt$$

If $V(t)$ is a random function of time, then Fourier analysis of samples of the signal of duration T taken at different times will lead to an ensemble of values for the coefficients $\{v_n\}$ and these, too, will be random variables. In the limit of large times T they are, moreover, uncorrelated for

$$\lim_{T\to\infty} T \left\langle v_n v_m^* \right\rangle = \lim_{T\to\infty} \frac{1}{T} \int_{-T/2}^{T/2} dt \int_{-T/2}^{T/2} dt' G(t-t') \exp\left[i(n-m)\omega t + m\omega(t-t')\right]$$

$$= \delta_{nm} \int_{-\infty}^{\infty} G(\tau) \exp(in\omega\tau) d\tau$$

(1.24)

When n = m we obtain the Wiener-Khinchin theorem

$$S(\omega) = \lim_{T\to\infty} \left\langle \frac{1}{T} \left| \int_{-T/2}^{T/2} V(t) \exp(i\omega t) dt \right|^2 \right\rangle = \int_{-\infty}^{\infty} G(\tau) \exp(i\omega\tau) d\tau \qquad (1.25)$$

which relates the power spectral density $S(\omega)$ of the signal to the autocorrelation function $G(\tau)$. Thus the power spectral density and the autocorrelation function are a Fourier transform pair. For example, if $V(t)$ has a Lorentzian power spectrum of half width at half height Γ, then the autocorrelation function is a negative exponential of decay, or correlation, time $\tau_c = 1/\Gamma$ (Figure 1.4). In other words, the time constant of the autocorrelation function (that reflects the fluctuation time of the signal) is equal to the inverse spectral bandwidth.

1.5   The Karhunen-Loeve Expansion

The transform pair of Equation 1.23 can be **generalized** to obtain an expansion in terms of orthogonal functions with uncorrelated coefficients even

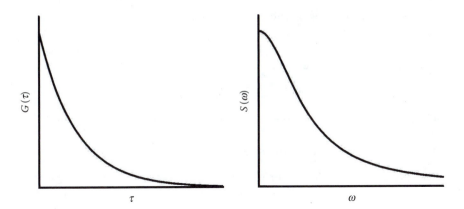

FIGURE 1.4
A Fourier transform pair of autocorrelation function and power spectral density.

when the interval T is finite (the *Karhunen-Loeve expansion*). For example, suppose the random function $V(t)$ is confined to the interval $[a,b]$

$$V(t) = \sum_n \sigma_n v_n \varphi_n(t) \quad a \le t \le b \tag{1.26}$$

where

$$\int_a^b \varphi_n(t)\varphi_m^*(t)dt = 1 \quad \text{if } m = n \tag{1.27}$$

$$= 0 \quad \text{if } m \ne n$$

$$\left\langle v_n v_m^* \right\rangle = 1 \quad \text{if } m = n \tag{1.28}$$

$$= 0 \quad \text{if } m \ne n$$

It is not difficult to show that

$$\left\langle V(t)V(s) \right\rangle = \sum_n |\sigma_n|^2 \varphi_n(t)\varphi_n^*(s) \quad a \le s,t \le b \tag{1.29}$$

The orthogonal functions satisfy the integral equation

$$\int_a^b \left\langle V(t)V(s) \right\rangle \varphi(s)ds = \lambda\varphi(t) \tag{1.30}$$

Here λ takes characteristic values equal to $|\sigma_n|^2$. This theorem is useful in calculations of the statistical properties of variables that have been averaged or *integrated* over fixed periods of time (see Chapter 13), although in practice the integral Equation (1.30) may be analytically intractable.

1.6   Statistical Independence

So far we have considered correlation between values of a single random variable at different times, **that is,** the *auto*correlation function of $V(t)$. However, we shall often be interested in the relationship between two or more random variables. For example, in the case of two variables, U and V, we can construct the single-time correlation function

$$G_{UV} = \langle U(t)V(t) \rangle = \iint UVP(U,V)dUdV \qquad (1.31)$$

If the two variables are uncorrelated then, by definition

$$\langle UV \rangle = \langle U \rangle \langle V \rangle \qquad (1.32)$$

Note that this does *not* imply that $\langle U^n V^m \rangle = \langle U^n \rangle \langle V^m \rangle$. In fact this more general property would be satisfied for all n and m only if U and V were *statistically independent*, that is, if

$$P(U,V) = P(U)P(V) \qquad (1.33)$$

1.7   Characteristic Functions and Generating Functions

A particularly useful aid in the theory of stochastic processes is the Fourier transform of the probability distribution: the *characteristic function*

$$C(\lambda) = \langle \exp(i\lambda V \rangle = \int_{-\infty}^{\infty} \exp(i\lambda V)P(V)dV \qquad (1.34)$$

The moments of the distribution can be obtained from this function by differentiation

$$\langle V^n \rangle = \left(-i\frac{d}{d\lambda}\right)^n C(\lambda)\Big|_{\lambda=0} \tag{1.35}$$

This procedure may be generalized to two or more variables. For example, the characteristic function for the joint distribution of U and V would be

$$C(\lambda,\mu) = \langle \exp\left[i(\mu U + \lambda V)\right]\rangle = \int_{-\infty}^{\infty} dU \int_{-\infty}^{\infty} dV P(U,V)\exp\left[i(\mu U + \lambda V)\right] \tag{1.36}$$

In the case of a sum of two statistically independent random variables, while the pdf of the resultant variable is a convolution of pdfs, the characteristic function is just a product

$$\langle \exp\left[i\lambda(U+V)\right]\rangle = \langle \exp(i\lambda U)\rangle\langle \exp(i\lambda V)\rangle \tag{1.37}$$

In the case of a variable that takes only positive values, it is convenient to use the Laplace transform of the distribution in a similar way. For example, for the variable $f = V^2$ we have

$$Q(s) = \langle \exp(-fs)\rangle = \int_{0}^{\infty} \exp(-sV^2)P(V)dV \tag{1.38}$$

This is known as the *generating function*. Moments of f may be generated from $Q(s)$ by evaluating its derivatives at $s = 0$. Generating functions for joint probability distributions may be defined by analogy with Equation 1.36 and its generalizations. These joint generating functions can be used to find moments and correlation functions.

1.8   Detection

The statistical formalism that has been presented so far can, in principle, be used to characterize and analyze data generated by any random process. However, the present book is devoted to the modelling of data generated by the scattering of electromagnetic and sound waves. This is acquired by an operation usually referred to as *detection* that limits the kind of variables and the **behavior** likely to be encountered in practice. In order to identify the statistical quantities of interest in the present context, therefore, we conclude this chapter with a brief discussion of detection systems.

In all remote sensing and communication systems, the received sound or electromagnetic wave is ultimately converted to an electrical signal and it is this postdetection signal that is subject to analysis and interpretation. Detection takes place in localized areas of space so that our postdetection signal is generally a function of spatial position as well as time $V(r,t)$. Many of the concepts reviewed in this chapter can be generalized to the spatial domain. As indicated earlier, there are obvious spatial analogs of a stationary ergodic random variable. The spatial correlation between values of a variable measured at two places often contains useful information, and the concept of correlation can be generalized to include both time differences and differences in position.

At the relatively long electromagnetic wavelengths relevant to radar, detectors respond to the electric field component of the electromagnetic wave. However, detectors used in the optical and infrared regions of the electromagnetic spectrum do not respond directly to the field (which oscillates on the order of 10^{14} hertz for visible light) but to the intensity, which is the power per unit area and is also the quantity to which the eye responds. So, for many statistical models, the fluctuating intensity is the quantity of interest.

This discussion has treated a classical wave. In the visible and near infrared parts of the electromagnetic spectrum, however, measurements at very low light levels may involve the quantum nature of light and employ *photon-counting* detectors. In this regime, the intensity I is replaced by a mean number of photo counts per unit area and statistical variations in the number of photo counts need to be taken into consideration when performing a measurement [5].

In practice, any detection system will have a finite collection area, finite response time, and finite dynamic range so that experimental data will usually be a spatial or time average of the incident signal and may also be amplitude limited. Moreover, experiments are always of finite duration so that measurements will be subject to statistical error. These effects will be considered in more detail in Chapter 13.

In the treatment of electromagnetic, and, to a lesser extent, acoustic, waves it is often possible to assume that the wave in question is *narrowband*. This means that the fundamental oscillation takes place on a much faster time scale than any other fluctuations (Figure 1.5)—an assumption that will be made for many of the calculations in this book. In such a regime, waves can be characterized by their amplitude and phase. An oscillating quantity such as the electric field component of an electromagnetic wave with amplitude A and phase ϕ may be written as

$$A(t)\cos\big(\omega t + \phi(t)\big) = X(t)\cos\big(\omega t\big) - Y(t)\sin\big(\omega t\big) \tag{1.39}$$

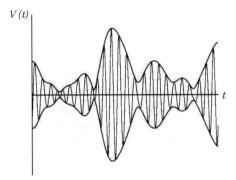

FIGURE 1.5
A narrowband random process.

and can be treated as narrowband if the bandwidths of the in-phase and quadrature components X and Y are much less than the oscillation frequency ω. It is then convenient to represent the field by a complex quantity or *phasor* E, that is

$$E(t) = A(t)\exp\left(i\phi(t)\right) = X(t) + iY(t) \tag{1.40}$$

The intensity is then proportional to the modulus squared of E

$$I(t) = \left|E(t)\right|^2 \tag{1.41}$$

with E expressed in appropriate units.

However, loss of information regarding the phase ϕ is not inevitable. The method known as *heterodyne* or *coherent* detection, which is also widely used in radar systems [4], can retain phase information. In this measurement method, the wave of interest is added to (mixed with) a second narrowband wave known as the local oscillator. The local oscillator has a frequency shift relative to the first wave, and the output of the detector contains a component that oscillates at the difference frequency. If the difference frequency (usually called the intermediate frequency) is $\Delta\omega$, the intensity illuminating the detector will be

$$\left|\sqrt{I}\exp\left(i\left[\phi + \omega t\right]\right) + \sqrt{I_{lo}}\exp\left(i\left[\phi_{lo} + (\omega - \Delta\omega)t\right]\right)\right|^2 = \tag{1.42}$$

$$I + I_{lo} + 2\sqrt{II_{lo}}\cos\left(\phi - \phi_{lo} + \Delta\omega t\right)$$

The frequency $\Delta\omega$ is chosen to be sufficiently high that the cosine term in Equation 1.42 varies much more rapidly than the other two terms. This means that a bandpass filter can be used to select out just this term, the result being a narrowband electrical signal that can be represented by the same phasor (except for an arbitrary constant phase shift) as the wave of interest. Note, however, that this filtering process will not completely remove the influence of the other terms. Quantum fluctuations in the measured intensities exist at all frequencies. These fluctuations, referred to as *shot noise*, inevitably pass through the filter and act as a fundamental noise source; when the local oscillator intensity exceeds a certain value, shot noise will dominate other sources of noise and maximum sensitivity will be achieved. Suitable demodulation of the narrowband electrical signal gives information about the fluctuations of the phase ϕ.

References

1. Davenport, W.B., and W.L. Root. *An Introduction to the Theory of Random Signals and Noise.* New York: IEEE Press, 1987.
2. Middleton, D. *An Introduction to Statistical Communication Theory.* New York: IEEE Press, 1996.
3. Goodman, J.W. *Statistical Optics.* New York: Wiley, 1985.
4. Skolnik, M.I. *Introduction to Radar Systems.* New York: McGraw-Hill, 1962.
5. Saleh, B. *Photoelectron Statistics.* New York: Springer-Verlag, 1978.

FIGURE 0.1
Random interference pattern generated when laser light is scattered by ground glass.

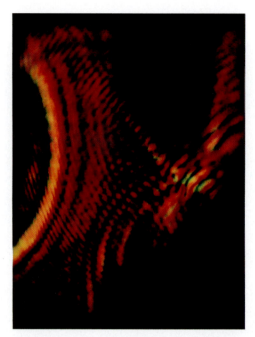

FIGURE 0.2
The effect of focusing multifrequency laser light onto a small area of the same kind of scatterer as in Figure 0.1.

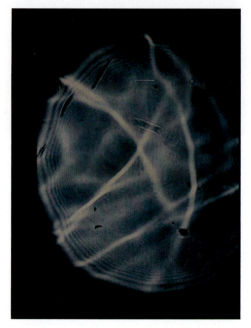

FIGURE 0.3
Intensity pattern obtained when laser light is passed through air convecting above a heating element.

2

The Gaussian Process

2.1 Introduction

The Gaussian process is by far the most widely studied stochastic model and has found many applications in both physical and biological sciences and in engineering. In the context of signal processing it is impossible to discuss Gaussian noise without mentioning the seminal papers of Rice [1], but his treatment of the subject added to an existing literature going back many decades. Papers of particular relevance to the later chapters of the present book date back to the beginning of the twentieth century and relate to certain population migration problems [2–4]. As a consequence of its long history, excellent accounts of this process are given in most standard textbooks on statistics and stochastic noise such as references 1–3 of Chapter 1. Here a summary will be presented of those features of the Gaussian process that are required to place in context the non-Gaussian models discussed later and to establish the range of their applicability to problems of practical interest.

2.2 Independent Gaussian Variables

Equation 1.14 defines the probability density characterizing a zero-mean Gaussian *variable*. The Fourier transform of this gives the corresponding characteristic function, which also has a Gaussian form

$$C(\lambda) = \exp\left(-\tfrac{1}{2}\lambda^2 \left\langle V^2 \right\rangle\right) \tag{2.1}$$

Here the average of the square of V is equal to the variance σ^2, since the mean is zero for this symmetric distribution. Equation 1.14 can be generalized to the case where $<V>$ is not zero by simply replacing V by $(V-<V>)$ in the exponents on the right-hand side of both Equation 1.14 and Equation 2.1. However, in the present book, zero-mean processes will be of most interest.

Equation 1.33 shows that the joint distribution of two statistically independent random variables is simply the product of the distributions of each one. In the case of two zero-mean Gaussian variables this leads to the formula

$$P(V_1, V_2) = \frac{1}{2\pi\sqrt{\langle V_1^2 \rangle \langle V_2^2 \rangle}} \exp\left(-\frac{V_1^2}{2\langle V_1^2 \rangle} - \frac{V_2^2}{2\langle V_2^2 \rangle}\right) \qquad (2.2)$$

The corresponding characteristic function is given by

$$C(\lambda_1, \lambda_2) = \exp\left[-\tfrac{1}{2}\left(\lambda_1^2 \langle V_1^2 \rangle + \lambda_2^2 \langle V_2^2 \rangle\right)\right] = C_1(\lambda_1)C_2(\lambda_2) \qquad (2.3)$$

These results are easily generalized to the case of variables with nonzero means as indicated above.

2.3   Correlated Gaussian Variables

Two *correlated* variables may be constructed by the linear transformation

$$U_1 = V_1 \cos\theta - V_2 \sin\theta$$
$$U_2 = V_1 \sin\theta + V_2 \cos\theta \qquad (2.4)$$

In the case of zero-mean variables, the joint probability density for these new variables can be obtained from Equation 2.2 using the method indicated by the relation shown in Equation 1.16 in Chapter 1. The corresponding characteristic function takes a particularly simple form and can be expressed in terms of the normalized correlation coefficient

$$g = \frac{\langle U_1 U_2 \rangle}{\sqrt{\langle U_1^2 \rangle \langle U_2^2 \rangle}} \qquad (2.5)$$

$$C_U(\lambda_1, \lambda_2) = \exp\left(-\tfrac{1}{2}\left\langle\left[\lambda_1 U_1 + \lambda_2 U_2\right]^2\right\rangle\right)$$

$$= \exp\left(-\tfrac{1}{2}\left[\lambda_1^2 + \lambda_2^2 + 2\lambda_1\lambda_2 g\right]\left\langle U^2\right\rangle\right) \tag{2.6}$$

The corresponding joint distribution is

$$P(U_1, U_2) = \frac{1}{2\pi\left\langle U^2\right\rangle\sqrt{1-g^2}} \exp\left[-\frac{U_1^2 + U_2^2 - 2gU_1U_2}{2\left\langle U^2\right\rangle(1-g^2)}\right] \tag{2.7}$$

It is clear from the transformation Equation 2.4, that because V_1 and V_2 are uncorrelated, $\left\langle U^2\right\rangle = \left\langle V_1^2\right\rangle + \left\langle V_2^2\right\rangle$. Note that Equation 2.6 has precisely the same form as Equation 2.1 but with $\tilde{V} = \lambda_1 U_1 + \lambda_2 U_2$ replacing λV. In fact, any linear combination of Gaussian variables is also a Gaussian variable.

It is now apparent from Equation 2.6 that there is a special relationship between correlation and statistical independence in the case of Gaussian variables, for if the correlation coefficient, g in this formula, is zero, the characteristic function factorizes as in Equation 2.3. Thus *two uncorrelated Gaussian variables are statistically independent.* This property is shared with some other random variables derived from the Gaussian, but is not generally true for non-Gaussian processes.

2.4 Higher-Order Correlations

Another important property revealed by Equation 2.6 is that the joint distribution of two Gaussian variables is a function only of the correlation, g, between the variables. In other words the higher-order correlations between two Gaussian variables can all be expressed in terms of the lowest order, g. Thus from Equation 2.6 it may be shown that

$$\frac{\left\langle U_1^{2n} U_2^{2m}\right\rangle}{\left\langle U_1^2\right\rangle^n \left\langle U_2^2\right\rangle^m} = \sum_{q=0}^{m}\binom{m}{q}\frac{\left(n+\tfrac{1}{2}\right)_q}{\left(\tfrac{1}{2}\right)_q} g^{2q}\left(1-g^2\right)^{m-q} \tag{2.8}$$

Here the mathematical symbols have the usually accepted interpretation in terms of factorials and gamma functions [5]:

$$\binom{m}{q} = \frac{m!}{q!(m-q)!}; \quad (\alpha)_q = \frac{\Gamma(\alpha+q)}{\Gamma(\alpha)} \tag{2.9}$$

As already indicated, all of the above results can be generalized to cases where the mean values of the two variables are nonzero. Results for the case of N correlated variables are also well documented in the literature. These are again most simply expressed in terms of the characteristic function $C(\{V_j\})$, for the multivariate distribution

$$C(\lambda_1, \lambda_2, \lambda_3, \ldots . \lambda_N) = \exp\left(-\tfrac{1}{2}\left\langle\left[\sum_{j=1}^{N} \lambda_j V_j\right]^2\right\rangle\right) \tag{2.10}$$

It can be seen that this is a simple generalization of Equation 2.6 and the corresponding distribution of zero-mean variables can be obtained by inverse Fourier transformation

$$P(V_1, V_2, \ldots V_N) = \frac{1}{(2\pi)^{N/2}|\Lambda|^{1/2}} \exp\left[-\frac{1}{2|\Lambda|}\sum_{n,m=1}^{N}|\Lambda|_{nm} V_n V_m\right] \tag{2.11}$$

Here Λ is the covariance matrix of elements $\lambda_{nm} = \langle V_n V_m \rangle$. All these results are valid for variables with unequal variances and are easily generalized to the case of variables with nonzero means.

2.5   Gaussian Processes

Consider now the properties of a random signal $V(t)$ that is a continuous function of time as discussed in Chapter 1. This quantity is said to be a *stationary Gaussian random process* if for every instant of time, t_j, the signal values $\{V(t_j)\}$ have the same joint Gaussian distribution (Figure 2.1). This means that the characteristic function has the form of Equation 2.10 but with the subscripts within the sum now referring to different instants of time, and with $\langle V_n^2 \rangle = \langle V^2 \rangle$. Probability densities can be derived by inverse Fourier transformation as before. For example, the two-time probability density has the same form as the two-variable case of Equation 2.7:

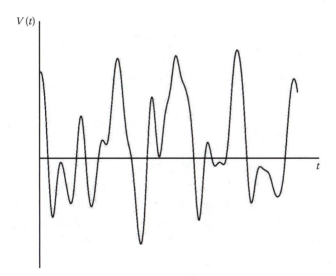

$V(t)$

t

FIGURE 2.1
A simulated joint Gaussian process.

$$P(V,V') = \frac{1}{2\pi\langle V^2\rangle\sqrt{1-g^2(\tau)}} \exp\left[-\frac{V^2 + V'^2 - 2VV'g(\tau)}{2\langle V^2\rangle\left((1-g^2(\tau))\right)}\right] \quad (2.12)$$

Here, the plain and primed values of V are deemed to be measurements at times t and $t + \tau$, respectively, and the correlation coefficient depends only on the time difference. It follows by analogy with Equation 2.11 that a Gaussian process is completely specified if the form of its correlation function $g(\tau)$ or equivalently its spectrum $S(\omega)$ (Equation 1.25) is known. This implies the existence of a hierarchy of factorization properties relating its higher-order correlations to g. For example, the fourth-order correlation may be expressed in terms of $g(\tau)$ as follows:

$$\frac{\langle V(t_1)V(t_2)V(t_3)V(t_4)\rangle}{\langle V^2\rangle^2} = \quad (2.13)$$

$$g(t_1 - t_2)g(t_3 - t_4) + g(t_1 - t_3)g(t_2 - t_4) + g(t_1 - t_4)g(t_2 - t_3)$$

This implies, for example, that $\langle V(0)^3 V(\tau)\rangle / \langle V^2\rangle^2 = 3g(\tau)$, and that $\langle V(0)^2 V(\tau)^2\rangle / \langle V^2\rangle^2 = 1 + 2g(\tau)^2$.
Another interesting result that can be derived using Equation 2.12 concerns the autocorrelation function of the rectangular wave $T(t)$ constructed with zeros that coincide with those of $V(t)$. Thus if $T(t) = +1$ for $V(t) > 0$ and $T(t) = -1$ if $V(t) < 0$ then

$$\langle T(0)T(\tau)\rangle = \frac{2}{\pi}\arcsin\left[g(\tau)\right] \qquad (2.14)$$

This is known as the arcsine formula or Van Vleck theorem [6].

2.6   Complex Gaussian Processes

It was indicated in Chapter 1 that the detected signal in a remote sensing system is most conveniently represented as a complex quantity, having both real and imaginary parts, or equivalently, in-phase and quadrature components. In this context it will be necessary to understand the properties of a *complex* Gaussian process

$$E(t) = X(t) + iY(t) \qquad (2.15)$$

The simplest case to consider is the *circular* complex Gaussian process when the real and imaginary parts $X(t)$ and $Y(t)$ are statistically independent zero-mean Gaussian processes with equal variance that are therefore uncorrelated at each instant of time, t. The joint distribution and joint characteristic function for these processes at a single time will be given by equations like Equation 2.2 and Equation 2.3. As indicated in Section 1.8, another way of representing Equation 2.15 is to use the amplitude A and phase ϕ in the complex plane

$$E(t) = A\exp(i\phi) \qquad (2.16)$$

Transformation from the variables X,Y to the new pair A,ϕ using Equation 2.16 leads to the joint density

$$P(A,\phi) = \frac{1}{\pi\langle A^2\rangle}A\exp\left(-\frac{A^2}{\langle A^2\rangle}\right) \quad \text{with} \quad 0<\phi<2\pi; \quad 0<A<\infty \qquad (2.17)$$

and the phase variable ϕ being uniformly distributed in the interval $[0,2\pi]$. Thus the amplitude probability density is a *Rayleigh distribution* (Figure 2.2)

$$P(A) = \frac{2A}{\langle A^2\rangle}\exp\left(-\frac{A^2}{\langle A^2\rangle}\right) \qquad (2.18)$$

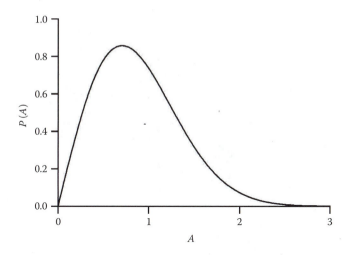

FIGURE 2.2
A Rayleigh distribution of unit mean.

In many remote sensing applications, in particular at optical frequencies, only the square of the amplitude will be measured. This quantity is usually referred to as the *intensity*, and is defined by

$$I(t) = |E(t)|^2 = A^2(t) = X^2(t) + Y^2(t) \qquad (2.19)$$

The probability density of the intensity fluctuations can be obtained by a further simple change of variable in Equation 2.18 or derived directly from the joint distribution (Equation 2.2) for X and Y

$$P(I) = \frac{1}{\langle I \rangle} \exp\left(-\frac{I}{\langle I \rangle}\right) \qquad (2.20)$$

Thus the intensity distribution for a circular complex Gaussian process is *negative exponential* (Figure 2.3) and the most likely outcome of a measurement of the intensity of such a process will be zero.

2.7 Joint Statistical Properties

The joint distribution (Equation 2.12) can be generalized to the case of a circular complex Gaussian process by using the characteristic function for four Gaussian variables. Although X and Y are uncorrelated, and therefore

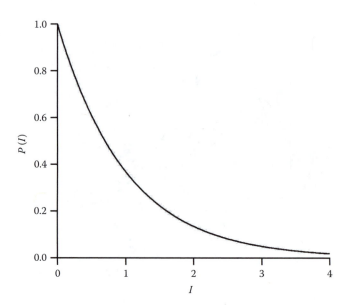

FIGURE 2.3
A negative exponential distribution.

statistically independent Gaussian variables at any given instant of time, there can be correlations between their values *at different times*. The joint distributions of amplitude and phase can be derived in a fairly straightforward manner

$$P(A, A') = \frac{4AA'}{\langle A^2 \rangle^2 [1 - |g^{(1)}(\tau)|^2]} I_0 \left(\frac{2AA' |g^{(1)}(\tau)|}{\langle A^2 \rangle [1 - |g^{(1)}(\tau)|^2]} \right)$$

$$\exp \left(-\frac{A^2 + A'^2}{\langle A^2 \rangle [1 - |g^{(1)}(\tau)|^2]} \right)$$

(2.21)

$$P(\phi, \phi') = \frac{1 - |g^{(1)}(\tau)|^2}{4\pi^2} \left[\frac{\sqrt{1 - \beta^2} + \beta(\pi - \cos^{-1} \beta)}{(1 - \beta^2)^{3/2}} \right] \quad when \quad 0 \le \phi, \phi' \le 2\pi$$

$$= 0 \quad otherwise$$

(2.22)

Here, $\langle X^2 \rangle = \langle Y^2 \rangle$, I_0 is the zero-order modified Bessel function [5] and

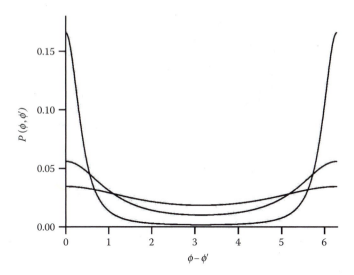

FIGURE 2.4
Joint distribution of the phase plotted as a function of phase difference for a circular complex Gaussian process.

$$g^{(1)}(\tau) = \frac{\langle EE'^* \rangle}{\langle |E|^2 \rangle} = \frac{\langle XX' \rangle}{\langle X^2 \rangle} + i \frac{\langle YX' \rangle}{\langle X^2 \rangle} = g(\tau) + ih(\tau) \qquad (2.23)$$

$$\beta = g(\tau)\cos(\phi - \phi') + h(\tau)\sin(\phi - \phi') \qquad (2.24)$$

The relationships $\langle XX' \rangle = \langle YY' \rangle$, $\langle XY' \rangle = -\langle YX' \rangle$ have been used to obtain these formulae. Note that when the spectrum of E is symmetric, its Fourier transform, the correlation function $g^{(1)}(\tau) = g(\tau)$, is real and the cross-correlation functions $\langle XY' \rangle$ and $\langle YX' \rangle$ vanish. According to Equation 2.22, the joint distribution of phase depends only on the modulus of the phase difference (Figure 2.4).

As in the case of a real Gaussian process, all the higher-order correlation functions of $E(t)$ can, in principle, be expressed in terms of the first-order correlation function [7]. The most often quoted of these *factorization* properties relates to the second-order intensity correlation function. This can be derived from a further simple transformation of Equation 2.21 that obtains the joint distribution of intensities and leads to the formula

$$g^{(2)}(\tau) = \frac{\langle II' \rangle}{\langle I \rangle^2} = 1 + \left| g^{(1)}(\tau) \right|^2 \qquad (2.25)$$

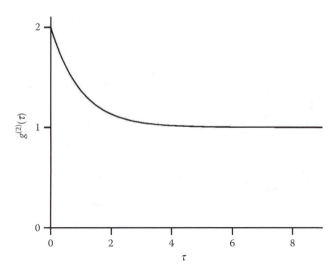

FIGURE 2.5
Intensity correlation function for a circular complex Gaussian process with a Lorentzian spectrum.

This is sometimes referred to as the Reed theorem [8], but is perhaps more correctly attributed to A. J. F. Siegert [9] who derived it in his wartime report on microwave scattering from raindrops. The behavior predicted for the correlation function of a complex Gaussian signal with symmetric Lorentzian spectrum is illustrated in Figure 2.5.

The phase correlation function can be derived from the joint distribution Equation 2.22, but one needs to be careful when considering the range of values taken by ϕ and ϕ'. First, note that in Equation 2.22 the values are defined to lie between 0 and 2π, but they may also be defined on any interval of length 2π. Freund and Kessler [10] have calculated a normalized phase correlation function, which is independent of which interval is chosen

$$C_{\phi\phi'} = \frac{\langle \phi\phi' \rangle - \langle \phi \rangle^2}{\langle \phi^2 \rangle - \langle \phi \rangle^2} \tag{2.26}$$

In the case of a symmetric spectrum ($h = 0$ in Equation 2.23), Freund and Kessler derive the following result

$$C_{\phi\phi'} = \frac{3}{2\pi^2} \left(\pi \sin^{-1}\left[g(\tau) \right] + \left(\sin^{-1}\left[g(\tau) \right] \right)^2 - \frac{1}{2} \sum_{n=1}^{\infty} \frac{\left[g(\tau) \right]^{2n}}{n^2} \right) \tag{2.27}$$

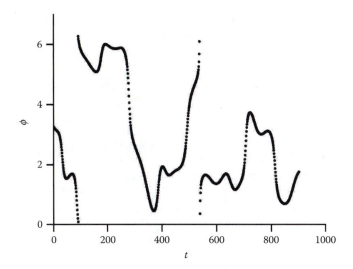

FIGURE 2.6
Simulation of a phase time series for a circular complex Gaussian process.

When ϕ is defined on a 0 to 2π interval, its mean is π and its mean squared value is $4\pi^2/3$; thus the phase correlation function is

$$\langle \phi\phi' \rangle = \pi^2 \left(1 - \frac{C_{\phi\phi'}}{3} \right) \tag{2.28}$$

The fact that the phase is confined to a fixed interval means that there are discontinuities in the phase time series; this can be seen in Figure 2.6, which shows results of numerical simulation. This means that the phase process behaves differently when ϕ is near 0 or 2π compared to when it is in the middle of the interval. This feature may be undesirable and sometimes it is preferable to avoid it by considering just the properties of the change in phase between times t and $t + \tau$. This can be visualized as the angle between the two vectors that represent the complex quantity E at the two different times. Note that the phase difference then only takes on values over a 2π range, whereas phase differences in Figure 2.6 go over a 4π range. A result for the mean square phase difference has been given by Tough et al [11]

$$\langle \Delta\phi^2 \rangle = \tfrac{1}{3}\pi^2 - \pi \sin^{-1}\left[g(\tau) \right] + \left(\sin^{-1}\left[g(\tau) \right] \right)^2 - \frac{1}{2} \sum_{n=1}^{\infty} \frac{\left| g^{(1)}(\tau) \right|^{2n}}{n^2} \tag{2.29}$$

where g and $g^{(1)}$ are defined by Equation 2.23. This is calculated for phase differences defined to lie between $-\pi$ and π. Equation (2.29) is calculated from the joint distribution (Equation 2.22), after multiplying by 2, by averaging the square of the difference variable using the identity

$$\int_{-\pi-\phi_0}^{\pi-\phi_0} d\phi \left(\sin^{-1}\left[\left| g^{(1)}(\tau) \right| \cos\phi \right] \right)^2 = \pi \sum_{n=1}^{\infty} \frac{\left| g^{(1)} \right|^{2n}}{n^2} \tag{2.30}$$

Donati and Martini also reported the result (Equation 2.29) in an earlier paper [12]. They noted that it was exactly half the result previously quoted by Middleton [13], which appears to have been integrated over a range that gives phase differences between 0 and 4π; note also that the phase autocorrelation function quoted in [13] is derived from the equation for the phase difference variance and is therefore not equivalent to Equation 2.27, and does not apply to the process illustrated in Figure 2.6.

Discontinuities in the phase seem unavoidable when starting from the two-time joint density. In practice, one often wants to deal with an unwrapped phase in which ϕ is continually extended to new 2π ranges. Calculations of properties of unwrapped phase can be performed by starting with a multiple-time joint density, as in the case of the phase derivative considered in Section 2.9 below. Indeed, the unwrapped phase is often defined as a time integral over the phase derivative [14.]

2.8 Properties of the Derivative of a Complex Gaussian Process

In some remote sensing systems and in communication systems, useful information is often manifest in properties of the *rate of change* of the signal rather than in the signal fluctuations themselves. Although there are idealized mathematical models for which the first or higher derivatives do not exist (see Chapters 5 and 6), in practice it is always possible to construct a quantity from the data, be it real or simulated, which approximates the derivative. Consider, for example the first difference

$$\dot{V}_\Delta(t) = \frac{1}{\Delta}\left[V(t+\Delta) - V(t) \right] \tag{2.31}$$

As Δ approaches zero, this expression approaches the first derivative of V provided that the function is continuous and once differentiable. If V is a Gaussian stochastic process, then the joint characteristic function for values of Equation 2.31 can be written down at different times

$$C\left(\{\lambda_j\}\right) = \left\langle \exp\left[i\sum_j \lambda_j \dot{V}_\Delta(t_j)\right]\right\rangle = \exp\left[-\frac{1}{2}\left\langle\left(\sum_j \lambda_j \dot{V}_\Delta(t_j)\right)^2\right\rangle\right] \quad (2.32)$$

The structure of this formula with respect to $\dot{V}_\Delta$ is identical to the characteristic function (Equation 2.10) for a Gaussian process. By taking the limit $\Delta \to 0$, therefore, it can be seen that if the derivative of V exists, it is also a Gaussian process. Now for any stationary and ergodic random process

$$\frac{d}{dt}\langle V(t)V(t)\rangle = 2\left\langle \frac{dV}{dt}V\right\rangle = 0 \quad (2.33)$$

Thus a random process and its derivative are uncorrelated. Therefore *a Gaussian process and its derivative are statistically independent*. Again, it must be emphasized that this property is only true for values at the same instant of time. $V(t)$ and its time derivative at another time $\dot{V}(t + \tau)$ will generally be correlated. By a small extension of the approach used in Equation 2.33, it may also be shown that for a continuous stationary ergodic process

$$\left\langle \dot{V}(t)\dot{V}(t+\tau)\right\rangle = -\frac{d^2}{d\tau^2}\left\langle V(t)V(t+\tau)\right\rangle = -\left\langle V^2\right\rangle \ddot{g}(\tau)$$

$$\left\langle \dot{V}^2\right\rangle = -\left\langle V^2\right\rangle \ddot{g}(0) = -\left\langle V^2\right\rangle / \tau_c^2 \quad (2.34)$$

These results can be used to calculate properties of the circular complex Gaussian process (Equation 2.15). The distribution of the rate of change of the amplitude appearing in Equation 2.16 and the intensity in Equation 2.19 are most easily obtained through a calculation of the relevant characteristic functions

$$\left\langle \exp(i\lambda\dot{A})\right\rangle = \left\langle \exp\left(i\lambda\frac{X\dot{X}+Y\dot{Y}}{\sqrt{X^2+Y^2}}\right)\right\rangle = \exp\left(-\frac{\lambda^2\left\langle\dot{X}^2\right\rangle}{2}\right) \quad (2.35)$$

$$\left\langle \exp(i\lambda\dot{I})\right\rangle = \left\langle \exp\left[-2\lambda^2\left(X^2+Y^2\right)\left\langle\dot{X}^2\right\rangle\right]\right\rangle = \left(1+4\lambda^2\left\langle X^2\right\rangle\left\langle\dot{X}^2\right\rangle\right)^{-1} \quad (2.36)$$

Here advantage has been taken of the statistical independence of $X, Y, \dot{X}, \dot{Y}$, and Equation 2.3 valid for independent Gaussian variables. The result (Equation 2.35) shows that the amplitude derivative of a circular complex Gaussian process is a zero-mean Gaussian variable. Inverse Fourier transformation of

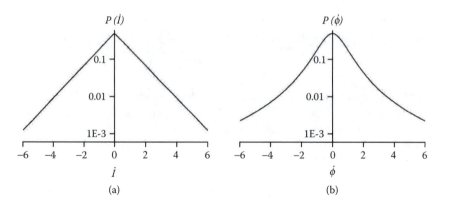

FIGURE 2.7
Circular complex Gaussian process: (a) log plot of the pdf of the intensity derivative, (b) log plot of the pdf of the phase derivative.

Equation 2.36 proves that fluctuations in the derivative of the intensity of such a process are governed by a two-sided exponential distribution (Figure 2.7). The distribution of the phase derivative can be calculated in a similar manner

$$\left\langle \exp(i\lambda\dot{\phi}) \right\rangle = \left\langle \exp\left(i\lambda \frac{X\dot{Y} - Y\dot{X}}{X^2 + Y^2} \right) \right\rangle = \left\langle \exp\left(-\frac{1}{2} \frac{\lambda^2 \left\langle \dot{X}^2 \right\rangle}{X^2 + Y^2} \right) \right\rangle =$$

$$2\lambda\sqrt{\left\langle X^2 \right\rangle \left\langle \dot{X}^2 \right\rangle} K_1\left(\lambda \sqrt{\frac{\left\langle \dot{X}^2 \right\rangle}{\left\langle X^2 \right\rangle}} \right)$$

$$(2.37)$$

$K_1(x)$ is a modified Bessel function of the second kind [5]. Here $\dot{\phi}$ can be considered as the limit of the difference $\Delta\phi$ that appears in Equation 2.29, and is unaffected by the discontinuities shown in Figure 2.6. Fourier transformation leads to the probability density

$$P(\dot{\phi}) = \frac{\tau_c / 2}{\left(\tau_c^2 \dot{\phi}^2 + 1 \right)^{3/2}} \tag{2.38}$$

Here, $\tau_c^2 = -\left\langle X^2 \right\rangle / \left\langle \dot{X}^2 \right\rangle$ as defined in Equation 2.34. It is important to note that Equation 2.38 is a "heavy-tailed" distribution, exhibiting only a cubic decay at large values of $\dot{\phi}$. It has a finite mean, but its variance and higher moments are infinite. This stems from the fact that as the amplitude of the complex process E passes close to zero in the complex plane, rapid changes in phase may occur. The behavior of Equation 2.38 is illustrated in Figure 2.7 for comparison with the exponential decrease that typically characterizes the fluctuation distributions of the intensity and its derivative.

The single-point derivatives of a complex Gaussian process possess a number of other interesting and potentially useful properties. For example, both the amplitude derivative and the amplitude-weighted phase derivative are statistically independent from any other function, f, of the real and imaginary parts X,Y. To see this, consider the characteristic function

$$C(\lambda_1,\lambda_2) = \left\langle \exp\left[i\left(\lambda_1 \dot{A} + \lambda_2 f(X,Y)\right)\right]\right\rangle = \left\langle \exp\left[-\frac{\lambda_1^2 \left\langle \dot{X}^2 \right\rangle}{2} + i\lambda_2 f(X,Y)\right]\right\rangle$$

$$= \exp\left(-\lambda_1^2 \left\langle \dot{X}^2 \right\rangle / 2\right)\left\langle \exp\left[i\lambda_2 f(X,Y)\right]\right\rangle$$

(2.39)

Since the characteristic function factorizes, $\dot{A}$ must be statistically independent from $f(X,Y)$. Exactly the same result is obtained if $\dot{A}$ is replaced by $A\dot{\phi}$, and these properties can be used to simplify the calculation of the statistics of other variables. For example, the characteristic function for the distribution of the "flux"

$$J = I\dot{\phi} = A^2\dot{\phi}$$

(2.40)

is just the average $\left\langle \exp\left(-A^2\lambda^2 \left\langle \dot{X}^2 \right\rangle / 2\right)\right\rangle$ over the Rayleigh distribution (Equation 2.18) for A. This leads to the simple result

$$P(J) = \frac{\tau_c}{\langle I \rangle} \exp\left(-\frac{2|J|\tau_c}{\langle I \rangle}\right)$$

(2.41)

Thus the fluctuations in J are governed by a two-sided negative exponential distribution, similar to that of the intensity derivative shown in Figure 2.7.

2.9 Joint Phase Derivative Statistics

A number of useful results for the two-time statistics of the derivatives of the amplitude, intensity, and phase of a circular complex Gaussian process can be found in the literature. The derivation of these results is based on the observation that $\dot{X}$, $\dot{Y}$, X, Y, and their values $\dot{X}'$, $\dot{Y}'$, X', Y' at a different time, are all Gaussian variables. Note that the correlation functions of $\dot{A}$ and $\dot{I}$ can be evaluated quite simply using Equation 2.33 and results such as Equation 2.25. However, as originally pointed out by Rice [15], this simple method cannot be used for the evaluation of the correlation function of the phase derivative because the phase is confined to the interval $[0,2\pi]$ in Equation 2.22 and therefore suffers discontinuities at the edges of the interval. Therefore, this chapter will be concluded with an explicit calculation of the phase derivative correlation function for the case of a circular complex Gaussian signal with nonsymmetric spectrum, an analysis that has been published only recently [16].

The correlation function of the phase derivative may be expressed as follows in terms of the real and imaginary parts of the signal

$$\left\langle \dot{\phi}(0)\dot{\phi}(\tau) \right\rangle = \left\langle \frac{X\dot{Y} - Y\dot{X}}{X^2 + Y^2} \frac{X'\dot{Y}' - Y'\dot{X}'}{X'^2 + Y'^2} \right\rangle \tag{2.42}$$

where the prime denotes evaluation at the later time, τ. In order to evaluate this quantity, a knowledge of the joint distribution of X,Y and their derivatives at two different times is needed. The required average is most easily accomplished through the eightfold characteristic function

$$C(z_1.....z_8) = \exp[-\tfrac{1}{2}\left\langle (z_1 X + z_2 X' + z_3 Y + z_4 Y' + z_5 \dot{X} + z_6 \dot{X}' + z_7 \dot{Y} + z_8 \dot{Y}')^2 \right\rangle] \tag{2.43}$$

that follows from standard Gaussian noise theory

$$\left\langle \dot{\phi}(0)\dot{\phi}(\tau) \right\rangle = \frac{1}{(2\pi)^8} \int\limits_{-\infty}^{\infty} dX...d\dot{Y}' \int\limits_{-\infty}^{\infty} dz_1...dz_8 \dot{\phi}\, \dot{\phi}' C(z_1...z_8) \exp[-i(z_1 X + ...z_8 \dot{Y}')] \tag{2.44}$$

The procedure adopted by Rice [15] is now followed by assuming that the signal is a stationary, zero-mean complex Gaussian process but with a non-symmetric spectrum. The reader is referred to his original paper for details of the symmetric spectrum case. Here only the principal steps in the calculation will be indicated. Noting that time derivatives occur only in the numerator of Equation 2.42, the averages over these variables in Equation 2.44 can be accomplished by taking appropriate derivatives of C with respect to $z_5 ... z_8$ evaluated at zero. The X,X',Y,Y' integrals can then be performed exactly to give the reduced expression

$$-\frac{1}{(2\pi)^4}\int\limits_{-\infty}^{\infty}\frac{dz_1..dz_4}{(z_1^2+z_3^2)(z_2^2+z_4^2)}\times\{\ddot{g}(z_1z_2+z_3z_4)+\dot{g}^2(z_1z_4-z_2z_3)^2+\ddot{h}(z_1z_4-z_2z_3)+$$

$$+\dot{h}^2(z_1z_2+z_3z_4)^2+b_1^2[(z_1z_2+z_3z_4)^2+(z_1z_4-z_2z_3)^2]+$$

$$+b_1\dot{g}(z_2z_3-z_1z_4)(z_1^2+z_2^2+z_3^2+z_4^2)+b_1\dot{h}(z_1z_2+z_3z_4)(z_1^2+z_2^2+z_3^2+z_4^2)+$$

$$+2\dot{g}\dot{h}(z_1z_2+z_3z_4)(z_2z_3-z_1z_4)\}\times$$

$$\times\exp\{-\tfrac{1}{2}[b_0(z_1^2+z_2^2+z_3^2+z_4^2)+2g(z_1z_2+z_3z_4)+2h(z_1z_4-z_2z_3)]\}\}$$

$$(2.45)$$

where

$$b_0=\left\langle X^2\right\rangle=\left\langle Y^2\right\rangle,\quad b_1=\left\langle X\dot{Y}\right\rangle=\left\langle X\dot{Y}'\right\rangle=-\left\langle Y\dot{X}\right\rangle=-\left\langle Y\dot{X}'\right\rangle,$$

$$g=\left\langle XX'\right\rangle=\left\langle YY'\right\rangle,\quad h=\left\langle XY'\right\rangle=-\left\langle YX'\right\rangle$$

and the derivatives of g and h are carried out with respect to the difference variable. Note that every z-dependent factor in the numerator of Equation 2.45 can be expressed as derivatives of the reduced characteristic function C (the final exponential factor) with respect to either g, h, or b_0. For example, the coefficient of the $\dot{g}$ term is $-\partial C/\partial g$ while the coefficient of the term in $b_1\dot{h}$ is $2\partial^2 C/\partial b_0\partial g$. Thus Equation 2.44 can be expressed as a sum of derivatives with respect to g, h, and b_0 of

$$\frac{1}{(2\pi)^2}\int\limits_{-\infty}^{\infty}\frac{dz_1..dz_4 C(z_1,z_2,z_3,z_4)}{(z_1^2+z_3^2)(z_2^2+z_4^2)}$$

$$=\frac{1}{(4\pi)^2}\int\limits_0^{\infty}du\int\limits_0^{\infty}dv\int\limits_0^{2\pi}d\theta\int\limits_0^{2\pi}d\phi\int\limits_0^{\infty}RdR\int\limits_0^{\infty}R'dR'\times$$

$$\times\exp\{-\tfrac{1}{2}[(b_0+v)R^2+(b_0+u)R'^2+ \quad\quad (2.46)$$

$$+2gRR'\cos(\theta-\phi)-2hRR'\sin(\theta-\phi)]\}$$

$$=\frac{1}{4}\int\limits_0^{\infty}\frac{dudv}{(b_0+u)(b_0+v)-(g^2+h^2)}=\frac{1}{4}\int\limits_1^{\infty}\frac{du\ln u}{u-q}$$

where $q=(g^2+h^2)/b_0^2$. The integrals in Equation 2.45 can all be expressed in terms of derivatives with respect to q of the final expression in Equation 2.46. This gives the result

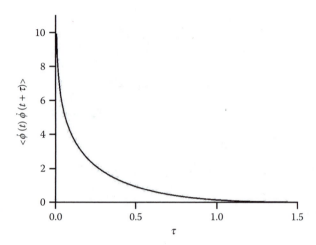

FIGURE 2.8
Circular complex Gaussian process: Phase derivative correlation function for the case of a symmetric Gaussian spectrum.

$$\left\langle \dot{\phi}(0)\dot{\phi}(\tau) \right\rangle = \left[\frac{g\ddot{g} + h\ddot{h} - \dot{g}^2 - \dot{h}^2}{2q} + \frac{(\dot{g}h - \dot{h}g)^2}{q^2} \right] \ln(1-q) +$$

$$+ \frac{b_1^2 + 2b_1(\dot{g}h - \dot{h}g) + (\dot{g}h - \dot{h}g)^2 / q}{(1-q)}$$

(2.47)

It is interesting to note that the coefficient of the *ln* term is a perfect derivative so that this term may be written

$$\frac{1}{4}\ln(1-q)\frac{d^2}{d\tau^2}\ln q$$

(2.48)

In the case of a symmetric spectrum, when b_1 and h are zero, Equation 2.47 reduces to that obtained by Rice [15]

$$\left\langle \dot{\phi}(t)\dot{\phi}(t+\tau) \right\rangle = \frac{1}{2}\left(\frac{g(\tau)\ddot{g}(\tau) - \dot{g}(\tau)^2}{g(\tau)^2} \right)\ln\left(1 - g(\tau)^2\right)$$

(2.49)

This is plotted in Figure 2.8. Note that the expression on the right-hand side of this formula is manifestly not proportional to the second derivative of either the phase correlation functions (Equation 2.27 and Equation 2.28), or the phase difference variance (Equation 2.29). Thus, as pointed out by Rice [15], the relationship in Equation 2.34 does not hold. This is because properties of the phase variable *modulo* 2 were used in the calculations of Section

2.7 while the phase derivative correlation function can be considered as a property of the *unwrapped* phase, requiring knowledge of the phase difference at two times, that is, a knowledge of the complex variable E at four times in all.

References

1. Rice, S.O. Mathematical Analysis of Random Noise. *Bell Syst. Tech. J.* 23, no. 24 (1944): 1–162; reproduced in *Noise and Stochastic Processes.* Ed. N. Wax. New York: Dover, 1954, 133–294.
2. Kluyver, J.C. A Local Probability Problem. *Proc. Roy. Acad. Sci. Amsterdam* 8 (1905): 341– 50.
3. Pearson, K. A Mathematical Theory of Random Migration. *Draper's Company Research Memoirs,* Biometric Series III, no. 15 (1906).
4. Lord Rayleigh (Strutt, J.W.). On the Problem of Random Vibrations in One, Two, or Three Dimensions. *Phil. Mag.,* 6 (1919): 321–47.
5. Abramowitz, M., and I.A. Stegun. *Handbook of Mathematical Functions.* New York: Dover, 1971.
6. Middleton, D. Some General Results in the Theory of Noise through Non-linear Devices. *Quart. Appl. Math.* 4 (1948): 445–98.; see also Chapter 13 in Reference [2] of Chapter 1.
7. Glauber, R.J. Coherent and Incoherent States of the Radiation Field. *Phys. Rev.* 131(1963): 2766–88.
8. Reed, I.S. On a Moment Theorem for Complex Gaussian Processes. *IRE Trans. Inf. Theory* IT-8 (1962): 194–95.
9. Siegert, A.J.F. *MIT Rad. Lab. Rep.* No. 463 (1943).
10. Freund, I., and D.A. Kessler. Phase Autocorrelation of Random Wave Fields. *Opt. Commun.* 124 (1996): 321–32.
11. Tough, R.J.A., D. Blacknell, and S. Quegan. A Statistical Description of Polarimetric and Interferometric Synthetic Aperture Radar Data. *Proc. Roy. Soc.* A 449 (1995): 567–89.
12. Donati, S., and G. Martini. Speckle-Pattern Intensity and Phase: Second-Order Conditional Statistics *J. Opt. Soc. Am.* 69 (1979): 1690–94.
13. Middleton, D. *An Introduction to Statistical Communication Theory.* New York: IEEE Press, 1996, Chapter 9, problem 9.4.
14. Roberts, J.H. *Angle Modulation.* Stevenage, UK: Peter Peregrinus, 1977, Chapter 3.
15. Rice, S.O. Statistical Properties of a Sine Wave Plus Random Noise. *Bell Syst. Tech. J.* 27 (1948): 109–57.
16. Ridley, K.D., and E. Jakeman. FM Demodulation in the Presence of Multiplicative and Additive Noise. *Inverse Problems* 15 (1999): 989–1002.

3

Processes Derived from Gaussian Noise

3.1 Introduction

Statistical models derived from Gaussian noise appear throughout the scientific literature and indeed are the subject of entire books [1]. This chapter reviews the properties of a number of processes based on Gaussian noise that are commonly encountered in modelling the statistics of scattered waves. Although these models have generally been developed with particular applications in mind, the context in which they have arisen will be covered in later chapters and here they will be discussed in abstract terms, with only their mathematical properties being presented. In anticipation of the applications of the models, general concern will be with the properties of narrowband complex processes that can be expressed in the form of Equation 1.40

$$E(t) = A(t) \exp[i\phi(t)] = X(t) + iY(t) \tag{3.1}$$

There are surprisingly few results in the literature relating to the phase statistics of such processes, and most of this chapter will be devoted to the description of intensity fluctuations. However, there is one important generalization of the Gaussian process for which results for the phase properties have been worked out, and this will be considered first.

3.2 Rice Variables

The simplest and most familiar generalization of the circular Gaussian *variable* is the sum of a complex variable E_G governed by such statistics and a second variable E_C with *constant* amplitude

$$E = E_G + E_C \tag{3.2}$$

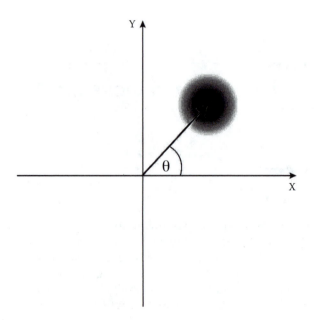

FIGURE 3.1
Constant vector plus Gaussian noise.

The resultant density function in the complex plane is a shifted Gaussian displaced by E_C and of width governed by the variance of the Gaussian noise component (see Figure 3.1). It is also possible to take the phase angle θ of E_C to be a random quantity, in which case, for uniform distribution of phase angle, the density function forms an annulus in the complex plane; in both cases the distributions of the amplitudes are given by Equation 3.4 below. From Equation 2.2 one obtains

$$P(X,Y) = \frac{1}{2\pi\langle X_G^2 \rangle} \exp\left[-\frac{(X - X_C)^2 + (Y - Y_C)^2}{2\langle X_G^2 \rangle} \right] \qquad (3.3)$$

This may also be written in terms of the amplitude and phase using the transformation of Equation 1.16

$$P(A,\phi) = \frac{A}{\pi\langle A_G^2 \rangle} \exp\left[-\frac{A^2 + A_C^2 - 2AA_C \cos(\phi - \theta)}{\langle A_G^2 \rangle} \right] \qquad (3.4)$$

where θ is the angle of the constant component. Integrating over the angle $\phi - \theta$ yields the so-called *Rice distribution* [2,3]

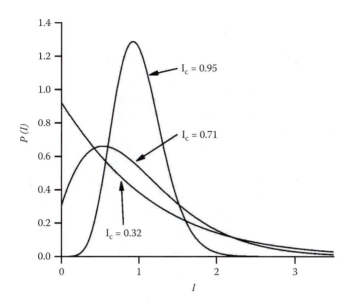

FIGURE 3.2
Distribution of the intensity of a Rice variable of unit mean for the values of I_C shown.

$$P(A) = \frac{2A}{\langle A_G^2 \rangle} \exp\left(-\frac{A^2 + A_C^2}{\langle A_G^2 \rangle} \right) I_0\left(2AA_C / \langle A_G^2 \rangle \right)$$

Here I_0 is a modified Bessel function of the first kind [4]. Defining the intensity of the variable as the square of the amplitude as in Equation 2.19, a simple transformation leads to

$$P(I) = \frac{1}{\langle I_G \rangle} \exp\left(-\frac{I + I_C}{\langle I_G \rangle} \right) I_0\left(2\sqrt{II_C} / \langle I_G \rangle \right) \tag{3.5}$$

This is plotted in Figure 3.2 for unit mean and some different values of I_c. The moments of this distribution can be obtained by direct integration or from the corresponding generating function. This is found also to be the discrete generating function for the class of Laguerre polynomials [4] and leads to the rather simple result (Figure 3.3)

$$\frac{\langle I^n \rangle}{\langle I \rangle^n} = n!\left(\frac{\langle I_G \rangle}{I_C + \langle I_G \rangle} \right)^n L_n\left(-I_C / \langle I_G \rangle \right) \tag{3.6}$$

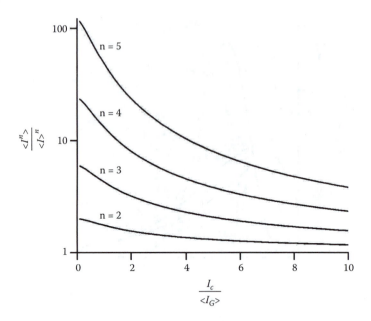

FIGURE 3.3
Normalized moments of a Rice variable as a function of I_C.

The second normalized moment takes the form

$$\frac{\langle I^2 \rangle}{\langle I \rangle^2} = 2 - \frac{I_C^2}{\langle I \rangle^2} \tag{3.7}$$

In Equation 3.6 and Equation 3.7 $\langle I \rangle = \langle I_C \rangle + \langle I_G \rangle$. Equation 3.7 demonstrates explicitly how the addition of a constant component to Gaussian noise tends to reduce the degree of fluctuation and provides a quantitative measure of the magnitude of the constant component. When $I_C = 0$ Equation 3.5 reduces to the negative exponential distribution (Equation 2.20) and only the $n!$ factor on the right-hand side of the result (Equation 3.6) survives, the second moment (Equation 3.7) being equal to two.

3.3 Rice Processes

If E_G is a circular complex Gaussian *process*, the higher-order joint statistical properties of the process (Equation 3.2) may be expressed entirely in terms of the correlation function $g^{(1)}(\tau)$ defined by Equation 2.23. Indeed the full joint probability density for the Rice *process* may be constructed straightforwardly

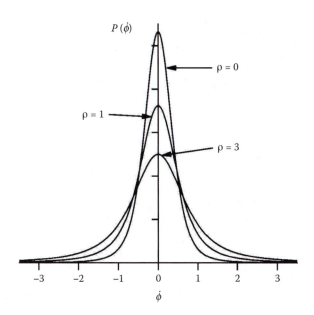

FIGURE 3.4
Distribution of the phase derivative of a Rice process, $\tau_c=1$.

from Equation 3.3 using the result from Equation 2.12. Since E_C is constant, the derivatives of the real and imaginary parts of a Rice process are Gaussian processes with properties governed by the results given in the previous chapter. Moreover, it is not difficult to show that the characteristic function for the amplitude derivative is given by Equation 2.35 as in the case of a simple complex Gaussian process. However, the distribution of the phase derivative of a Rice process, shown in Figure 3.4, differs from the previous result (Equation 2.38) and is

$$P(\dot{\phi}) = \frac{(\tau_c / 2)\exp(y / 2 - \rho)}{\left(\tau_c^2 \dot{\phi}^2 + 1\right)^{3/2}}\left[(y + 1)I_0(y / 2) + yI_1(y / 2)\right] \qquad (3.8)$$

where

$$\rho = I_C / \langle I_G \rangle, \quad y = \rho / (\tau_c^2 \dot{\phi}^2 + 1), \quad \tau_c^2 = \langle X^2 \rangle / \langle \dot{X}^2 \rangle$$

When the spectrum is symmetric (real $g(\tau)$) the correlation function of the phase derivative can also be obtained in terms of special functions using the method outlined in Section 2.9. The result may be expressed in the form [3]

$$\langle \dot{\phi}(0)\dot{\phi}(\tau) \rangle = \frac{1}{2} \frac{\dot{g}(\tau)^2 - g(\tau)\ddot{g}(\tau)}{g(\tau)^2} f_1 - \frac{\dot{g}(\tau)^2}{2g(\tau)^2} f_2 \qquad (3.9)$$

Here the functions f_1 and f_2 are given in terms of the exponential integral Ei [4] by

$$f_1 = \exp\left(\frac{I_C}{\langle I_G \rangle g(\tau)}\right)$$

$$\left\{ Ei\left[\frac{I_C}{\langle I_G g(\tau) \rangle}\right] - 2Ei\left[\frac{I_C(1 - g(\tau))}{\langle I_G \rangle g(\tau)}\right] + Ei\left[\frac{I_C(1 - g(\tau))}{\langle I_G \rangle g(\tau)(1 + g(\tau))}\right] \right\}$$

$$f_2 = \frac{I_C f_1}{\langle I_G \rangle g(\tau)} - 1 + \frac{2}{1 - g(\tau)} \exp\left(-\frac{I_C}{\langle I_G \rangle}\right) - \frac{1 + g(\tau)}{1 - g(\tau)} \exp\left(-\frac{2I_C}{\langle I_G \rangle(1 + g(\tau))}\right)$$

$$(3.10)$$

When $I_C = 0$, f_2 vanishes while $f_1 = \ln(1 - g(\tau)^2)$ so that Equation 3.9 reduces to the result of Equation 2.49 as expected.

3.4 Gamma Variables

Perhaps the most familiar intensity distribution that can be constructed from Gaussian variables is the *chi-square* or *m-distribution* [1,5]. This is obtained through the addition of the squares of m statistically independent but identically distributed Gaussian variables

$$I = \sum_{n=1}^{m} X_n^2 \qquad (3.11)$$

The distribution can be obtained without difficulty from the generating function and takes the form

$$P(I) = \frac{b^{m/2} I^{m/2 - 1}}{\Gamma(\frac{1}{2} m)} \exp(-bI) \qquad (3.12)$$

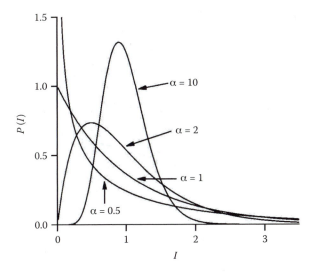

FIGURE 3.5
Gamma distributions of unit mean and second moments 1.1, 1.5, 2, and 3 (α = 10, 2, 1,and 0.5 as shown).

The negative exponential distribution, case m = 2, corresponds to the zero-mean circular complex Gaussian variable result (Equation 2.20) encountered in the previous chapter.

Equation 3.12 can be analytically continued to noninteger values of m, leading to the class of *gamma* variates plotted in Figure 3.5

$$P(I) = \frac{b^{\alpha} I^{\alpha-1}}{\Gamma(\alpha)} \exp(-bI) \tag{3.13}$$

The normalized moments of this two-parameter class of distributions are given by the following ratio of gamma functions

$$\frac{\langle I^n \rangle}{\langle I \rangle^n} = \frac{\Gamma(n+\alpha)}{\alpha^n \Gamma(\alpha)} \tag{3.14}$$

These are functions of only one parameter and can be expressed in terms of the second normalized moment

$$\frac{\langle I^2 \rangle}{\langle I \rangle^2} = 1 + \frac{1}{\alpha} \tag{3.15}$$

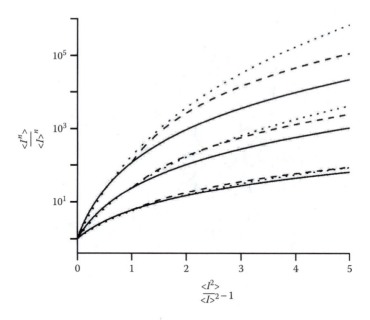

FIGURE 3.6
The normalized third, fourth, and fifth moments of gamma (solid line), K (dashed line) and lognormal (dotted line) distributions.

Figure 3.6 shows the third and higher moments of the gamma class plotted against the second moment—a set of curves that, in principle, provides a unique signature of the fluctuation properties of this type of variable.

As expected, Equation 3.15 takes the zero-mean circular complex Gaussian variable value of 2 when $\alpha = 1$ corresponding to the case when $m = 2$ in the definition (Equation 3.11). As α increases, the degree of fluctuation decreases. This is only to be expected since the fluctuations are then averaged out through the addition of more and more terms in the sum (Equation 3.11). However, when α is less than unity, the fluctuations governed by Equation 3.13 are greater than those of a complex Gaussian variable, and when α is less than one half, the fluctuations are greater than those of a real Gaussian variable.

3.5   Gamma Processes

Since the definition (Equation 3.11) expresses the m-distribution in terms of Gaussian variables, this model can easily be generalized to include time evolution through the use of stationary Gaussian *processes* in the sum. The generating function for the joint distribution of intensities at two different times is defined as a generalization of Equation 1.38

$$Q(s,s') = \langle \exp(-sI - s'I') \rangle \tag{3.16}$$

This can be calculated using the results of the last chapter and takes the rather simple form

$$Q(s,s') = \left\{ \left(1 + \frac{\langle I \rangle s}{\sqrt{\alpha}} \right) \left(1 + \frac{\langle I \rangle s'}{\sqrt{\alpha}} \right) - \frac{\langle I \rangle^2 ss'}{\alpha} \left| g^{(1)}(\tau) \right|^2 \right\}^{-\alpha} \tag{3.17}$$

Here, $\alpha = m/2$; however, the more general gamma process may be obtained by analytic continuation to include any positive value of α. The amplitude model obtained by setting $I = A^2$ is sometimes called the generalized Rayleigh process [6]. It will be seen in Chapter 12 that in the case of a Lorentzian spectrum, this kind of process can in fact be generated without any reference to underlying Gaussian statistics [7,8].

The joint density for the gamma process can be obtained by inverse Laplace transformation of the result in Equation 3.17, giving

$$P(I,I') = \frac{\alpha \left(\sqrt{\alpha II'} \big/ \langle I \rangle \left| g^{(1)}(\tau) \right| \right)^{\alpha-1}}{\langle I \rangle^2 \Gamma(\alpha) \left[1 - \left| g^{(1)}(\tau) \right|^2 \right]} \exp \left[\frac{-\sqrt{\alpha}(I + I')}{\langle I \rangle \left[1 - \left| g^{(1)}(\tau) \right|^2 \right]} \right] I_{\alpha-1} \left(\frac{2 \left| g^{(1)}(\tau) \right| \sqrt{\alpha II'}}{\langle I \rangle \left[1 - \left| g^{(1)} \right|^2 \right]} \right) \tag{3.18}$$

Recalling that $I = A^2$, the result in Equation 2.21 for the complex Gaussian case, $m = 2$, is recovered on setting $\alpha = 1$. Just as in the case of a complex Gaussian process, the result in Equation 3.18 implies certain factorization properties that are manifest in the higher-order moments and correlation functions. The simplest and most frequently used of these is a generalization of the Siegert relation (Equation 2.25) that expresses the intensity correlation function in terms of the spectrum of the underlying Gaussian variables

$$\frac{\langle II' \rangle}{\langle I \rangle^2} = 1 + \frac{1}{\alpha} \left| g^{(1)}(\tau) \right|^2 \tag{3.19}$$

Some higher-order factorization properties have also found applications and may be calculated from the definition in Equation 3.11. In the case of a symmetric spectrum ($g^{(1)}(\tau) = g(\tau)$), for example, the triple correlation function is given by

$$\frac{\langle I_1 I_2 I_3 \rangle}{\langle I \rangle^3} = 1 + \frac{1}{\alpha}\left(g_{12}^2 + g_{23}^2 + g_{31}^2\right) + \frac{2}{\alpha^2} g_{12} g_{23} g_{31} \tag{3.20}$$

Here, the subscripts refer to different times with $g_{ij} = g(t_i - t_j)$.

3.6 Statistics of the Derivative of a Gamma Process

The distribution of the time derivative of a gamma process may be deduced by setting $s = i\lambda/\tau, s' = i\lambda/\tau$ in the result of Equation 3.17. This allows the characteristic function of an intensity difference variable to be written down immediately

$$C(\lambda;\tau) = \left\langle \exp\left(\frac{i\lambda}{\tau}\left[I(t) - I(t+\tau)\right]\right)\right\rangle = \left\{1 + \frac{\langle I \rangle^2 \lambda^2}{\alpha^2 \tau^2}\left[1 - \left|g^{(1)}(\tau)\right|^2\right]\right\}^{-\alpha} \tag{3.21}$$

In the limit of small τ the exponent reduces to the time derivative of the intensity, assuming this is a differentiable process. The following result is then obtained (recalling that $g^{(1)} = g + ih$, Equation 2.23)

$$C(\lambda;0) = \left\{1 + \frac{\langle I \rangle^2 \lambda^2 [\ddot{g}(0) + \dot{h}^2(0)]}{\alpha^2}\right\}^{-\alpha} \tag{3.22}$$

For example, in the case of a real Gaussian correlation function such that $h(t) = 0$ and $g^{(1)}(t) = g(t) = \exp(-t^2/\tau_c^2)$, the product $\langle I \rangle^2 \ddot{g}(0)$ takes the value $2\langle I \rangle^2/\tau_c^2$ and can be interpreted as a mean square intensity gradient, $\langle \dot{I}^2 \rangle$ (*cf.* result in Equation 2.34).

The inverse Fourier transform of the result in Equation 3.22 is the probability density function of the intensity derivative and may be expressed in terms of modified Bessel functions of the second kind [4]:

$$P(\dot{I}) = \frac{2^{1-\alpha}|\dot{I}|^{\alpha-\frac{1}{2}}}{\Gamma(\alpha)} \left(\frac{\alpha}{\langle \dot{I}^2 \rangle} \right)^{(\alpha+1/2)/2} K_{\alpha-1/2}\left(|\dot{I}| \sqrt{\frac{\alpha}{\langle \dot{I}^2 \rangle}} \right) \tag{3.23}$$

Distributions based on K Bessel functions (also known as McDonald functions) can arise in a number of different ways and have found wide application in scattering theory for reasons that will be discussed in later chapters. Another important kind of model that can lead to a result of the type seen in Equation 3.23 will be described next. This is a *compound* process in which the local mean of one noise process is modulated by a second one that may be fluctuating on a different time scale.

3.7 Compound Variables

In the present context, the simplest compound variable is obtained through the modulation of the average intensity of one gamma variable by another [1]. The single point statistics are calculated by simply averaging the distribution (Equation 3.13) over fluctuations in $\langle I \rangle$ governed by the same distribution with, for example, parameter β. This gives

$$P(I) = \frac{2b}{\Gamma(\alpha)\Gamma(\beta)} (bI)^{\frac{\alpha+\beta}{2}-1} K_{\alpha-\beta}\left(2\sqrt{bI} \right) \tag{3.24}$$

It is clear by taking $\beta = {}^1\!/_2$ and setting $I \propto \dot{I}^2$ that the distribution (Equation 3.23) belongs to the same class as Equation 3.24 although the origin of the fluctuations is quite different.

A particularly important class of two-parameter distributions, known as K-distributions [9], is obtained when $\beta = 1$

$$P(I) = \frac{2b}{\Gamma(\alpha)} (bI)^{\frac{\alpha-1}{2}} K_{\alpha-1}\left(2\sqrt{bI} \right) \tag{3.25}$$

The result in Equation 3.25 will be adopted as the standard form for the class of K-distributions and the associated processes referred to as K-*distributed noise* [10]. Some examples of the probability density function in Equation 3.25 are plotted in Figure 3.7. Note that, unlike some of the other distributions discussed in this chapter, the pdf always goes to a nonzero value as I goes to zero (a finite value of $b/(\alpha - 1)$ for $\alpha > 1$, and infinity for $\alpha \leq 1$). Distributions of the type in Equation 3.25 arise when the mean intensity of a

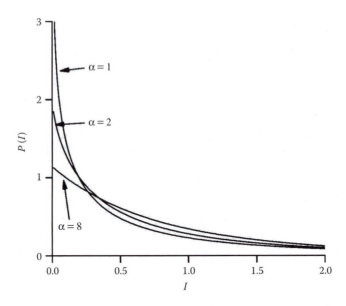

FIGURE 3.7
K-distributions of unit mean, with second moments of 2.25, 3, and 4 (α = 8, 2, and 1 as shown).

complex Gaussian variable is itself a gamma variate. Therefore the fluctuations described by K-distributions always exceed those of the intensity of a complex Gaussian variable, and their moments can always be expressed as the product of those of a gamma variate with those of an exponential variate. Thus it follows from Equation 3.14 that the normalized moments are given by

$$\frac{\langle I^n \rangle}{\langle I \rangle^n} = n! \frac{\Gamma(n+\alpha)}{\alpha^n \Gamma(\alpha)} \tag{3.26}$$

The higher moments are all parameterized by the second and are plotted in Figure 3.6 for comparison with those of a gamma variate. Equation 3.26 is just the product of the normalized moments of negative exponential and gamma variables, and is therefore greater than either one.

The two-point statistics of K-distributed noise are model dependent and will be discussed in Chapter 12 together with the properties of a more general compound distribution involving the K Bessel functions. The interplay of time scales between the two processes underlying this kind of noise is subtle, and few results of a general nature have been published.

3.8 Other Commonly Encountered Distributions

An infinite range of new probability densities can be generated by constructing *functions* of Gaussian variables. A few of these have been used to model fluctuations in scattered waves, the most widely known being the lognormal distribution [11]. This is the probability density of a variable whose logarithm is Gaussian distributed. The variable may thus be written in the form

$$\psi = C \exp(X) \tag{3.27}$$

where $P(X) = \exp(-X^2/2\sigma^2)/\sqrt{2\pi\sigma^2}$ and C is a constant. Applying the transformation rule (Equation 1.11) gives

$$P(\psi) = \frac{1}{\psi\sqrt{2\pi\sigma^2}} \exp\left[-\frac{\left(\ln\psi - \ln C\right)^2}{2\sigma^2}\right] \tag{3.28}$$

Equation 3.27 defines a quantity evolving in time when X is assumed to be a Gaussian process. The moments and correlation functions can be deduced from Equation 3.27 using the property of Equation 2.10. For the normalized moments one obtains

$$\frac{\left\langle\psi^n\right\rangle}{\left\langle\psi\right\rangle^n} = \exp\left[\tfrac{1}{2}n(n-1)\sigma^2\right] \tag{3.29}$$

These moments grow dramatically with order even for moderate values of σ, reflecting the heavy tail of the lognormal distribution.

The correlation function of ψ may be derived directly from Equation 3.27 with the help of Equation 2.10

$$\frac{\left\langle\psi\psi'\right\rangle}{\left\langle\psi\right\rangle^2} = \exp\left[\sigma^2 g(\tau)\right] \tag{3.30}$$

Here g is the correlation function of the real Gaussian process X.

The definition of Equation 3.27 can be generalized to the case of a complex Gaussian process

$$\psi = C \exp(X + iY) \tag{3.31}$$

Moments of the intensity $I = |\psi|^2$ are given by Equation 3.29 with even values of n and may again be very large unless the variance of the underlying Gaussian process X is small. However, moments of ψ itself, as opposed to its modulus, may be small even if the variances of X and Y are both large. For example, if the underlying processes are statistically independent

$$\frac{\langle \psi^n \rangle}{\langle \psi \rangle^n} = \exp\left[\tfrac{1}{2}n(n-1)(\sigma_X^2 - \sigma_Y^2)\right] \tag{3.32}$$

which *decreases* with order if the variance of Y exceeds that of X. The correlation function of the quantity in Equation 3.31 when X and Y are Gaussian processes can also be calculated while the full joint distribution of ψ may be obtained using the Gaussian result of Equation 2.10 and the general transformation expressed in Equation 1.16.

Moments of I are plotted in Figure 3.6, which compares the moments of lognormal, K, and gamma distributions having the same second normalized moment. Note that, unlike the case of the K-distribution, the second normalized moment of the lognormal and gamma distributions may be less than two. The pdf of I is the same as Equation 3.28, with C replaced by C^2 and σ^2 by $4\sigma^2$ and the average value of I is $C^2 \exp(2\sigma^2)$. Examples of the pdf are plotted in Figure 3.8. Note that $P(I)$ always goes to zero as I goes to zero, but the peak in the density function moves to smaller values of I as the second moment increases.

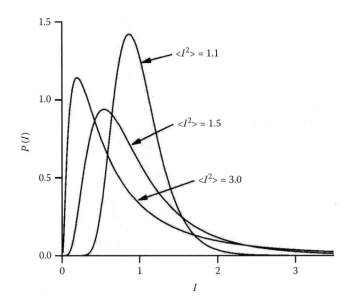

FIGURE 3.8
Lognormal distribution of unit mean for second moments shown.

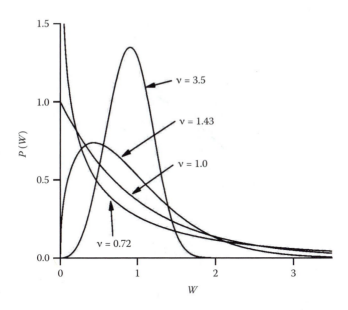

FIGURE 3.9
Weibull distribution of unit mean for second moments of 1.1, 1.5, 2, and 3 (v = 3.50, 1.43, 1.0, and 0.72 as shown).

As a final example, consider applying the transformation $I = W^v$ to the negative exponential distribution in Equation 2.20, recalling that this distribution is obtained when the variable I is the modulus of a circular complex Gaussian process. It follows from Equation 1.10 that

$$P(W) = \frac{vW^{v-1}}{\langle I \rangle} \exp\left(-\frac{W^v}{\langle I \rangle} \right) \tag{3.33}$$

This is the density corresponding to the *Weibull* distribution [12] that has found application in a number of fields, including wave scattering. This density function is shown in Figure 3.9. The normalized moments of this distribution are given by

$$\frac{\langle W^n \rangle}{\langle W \rangle^n} = \frac{\Gamma(1+n/v)}{\Gamma^n(1+1/v)} \tag{3.34}$$

Weibull densities are similar to the class of K-distributions in Equation 3.25 for some parameter values, and comparison of Equation 3.26 and Equation 3.34 shows that members of the two classes coincide when $v = \alpha = 1/2$. The normalized moments of the distributions are compared in Figure 3.10. A

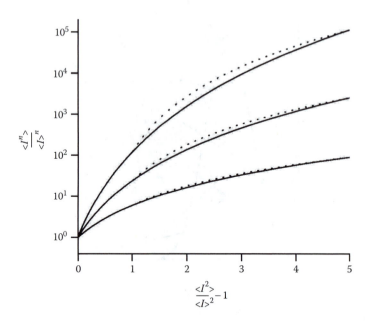

FIGURE 3.10
Normalized moments of the K (dashed line) and Weibull (solid line) distributions.

comparison of Figure 3.9 and Figure 3.5 also indicates the similarities between the Weibull and Gamma distributions.

Equation 3.34 effectively predicts the *fractional* moments of the sum of the squares of two identical zero-mean Gaussian variables. It is possible to define the equivalent fractional *process* $W = \left(X^2 + Y^2\right)^{1/\nu}$ where X and Y are identical zero-mean Gaussian processes. The correlation function of W may then be calculated from Equation 3.18 with $\alpha = 1$

$$\frac{\langle W(0)W(t)\rangle}{\langle W\rangle^2} = \left(1 - \left|g^{(1)}(t)\right|^2\right)^{1+2/\nu} F\left(1+\tfrac{1}{\nu}, 1+\tfrac{1}{\nu}; 1; \left|g^{(1)}(t)\right|^4\right) \qquad (3.35)$$

Note that the hypergeometric function in this formula may be expressed in terms of Legendre functions [4]. As the Weibull distribution is often used in situations where the underlying phenomenology leading to its relevance is unclear, it is perhaps not surprising that time-dependent processes such as that leading to Equation 3.35 are not discussed in the literature.

Finally, it should be mentioned that Roberts [13] has attempted to develop a theory for non-Gaussian processes that includes the phase difference variable $\Delta = \phi - \phi'$ (as in Section 2.7 for the Gaussian process) in a form for the joint distribution of amplitudes at two different times. In particular he showed that some development was possible for the case of the Weibull/K-distributions of index $^1/_2$. However, the analytical results generally appear to be of only limited utility.

References

1. Miller, K.S. *Multidimensional Gaussian Distributions*. The SIAM Series in Applied Mathematics. Eds. R.F. Drenick and H. Hochstadt. New York: John Wiley, 1964.
2. Rice, S.O. Mathematical Analysis of Random Noise. *Bell Syst. Tech. J.* 23, no. 24, (1944): 1–162 ; reproduced in *Noise and Stochastic Processes*. Ed. N. Wax. New York: Dover, 1954, 133–294.
3. Rice, S.O. Statistical Properties of a Sine Wave Plus Random Noise. *Bell Syst. Tech. J.* 27 (1948): 109–57.
4. Abramowitz, M., and I.A. Stegun. *Handbook of Mathematical Functions*. New York: Dover, 1971.
5. Nakagami, M. "The m-Distribution — A General Formula of Intensity Distribution of Rapid Fading." In *Statistical Methods in Radio Wave Propagation*. Ed. W.C. Hoffman. New York: Pergamon, 1960.
6. Stratonovich, R.L. *Topics in the Theory of Random Noise*, Vol 1. New York: Gordon and Breach, 1963.
7. Wong, E. The Construction of a Class of Stationary Markoff Processes. *Proc. Am. Math. Soc. Symp. Appl. Math.* 16, (1963): 264–76.
8. Jakeman, E. Statistics of Integrated Gamma-Lorentzian Intensity Fluctuations. *Opt. Acta* 27 (1980): 735–41.
9. Jakeman, E., and P.N. Pusey. A Model for Non-Rayleigh Sea Echo. *IEEE Trans. Antenn. Propag.* AP-24 (1976): 806–14.
10. Jakeman, E. On the Statistics of K-Distributed Noise. *J. Phys. A: Math. Gen.* 13 (1980): 31–48.
11. Aitchison, J. *The Lognormal Distribution with Special Reference to its Uses in Economics*. Cambridge: Cambridge University Press, 1963.
12. Weibull, W. A Statistical Distribution Function of Wide Applicability. *J. Appl. Mech.* 18 (1951): 293–97.
13. Roberts, J.H. Joint Intensity and Envelope Densities of a Higher Order. *IEE Proc. Sonar Navig.* 142 (1995): 123–29.

4

Scattering by a Collection of Discrete Objects: The Random Walk Model

4.1   Introduction

In this chapter we discuss a scattering problem that can be characterized by a simple but extremely powerful mathematical model: the random walk on a plane. This model has a wide range of applications throughout science and engineering and many of its statistical properties were derived early in the last century [1–3]. In the present context it provides a description of scattering by a finite collection of discrete objects. It is an essentially exact model for light scattering by small particles, and in this chapter we shall often derive results and use experimental illustrations based on such systems (see, for example, References 4 and 5). However, the model is equally relevant to microwave scattering from raindrops or electron scattering from atomic defects, and it will be demonstrated in later chapters that it also provides a good representation for many aspects of scattering by continuous systems, such as rough surfaces and turbulent media.

We shall restrict consideration in the present chapter to the scattering of monochromatic scalar waves such as sound waves or a single polarized component of electromagnetic radiation, and we shall also assume that the collection of scatterers is sufficiently dilute that multiple scattering can be neglected. The effect of broadband illumination, fluctuations in the polar-ization of scattered vector waves, and some multiple scattering effects will be considered in later chapters.

4.2   The Incident Wave

As discussed at the end of Chapter 1, a quasi-monochromatic incident wave can be written in complex representation as

$$E_0(\bar{r},t) = A_0(\bar{r},t)\exp\left[i\left(\varphi_0(\bar{r},t)+kz-\omega t\right)\right] \qquad (4.1)$$

where ω is the angular frequency, k is the magnitude of the wave vector, and z is the propagation direction. The intensity of the illumination is the square of the modulus of the field E_0

$$I_0 = |E_0|^2 = A_0^2 \qquad (4.2)$$

and will here be taken to be time independent. The simplest case to consider is that for which the incident intensity is spatially uniform throughout the scattering region. However, there will often be some spatial variation of intensity, and in laser light scattering a TEM00 Gaussian is usually employed; this has the following intensity profile [6]

$$I_0(\bar{r},t) = I_0 \exp\left(-\frac{2R^2}{W(z)^2}\right) \qquad (4.3)$$

Here we have adopted the coordinate system $\bar{r} \equiv (\bar{R}, z)$. The parameter W is the beam radius, and this varies with the propagation distance z. For most purposes, the following paraxial formula provides an excellent approximation

$$W^2(z) = W_0^2\left[1+\left(\frac{2z}{kW_0^2}\right)^2\right] \qquad (4.4)$$

W_0 is the minimum value of the beam radius, which occurs at $z = 0$. The phase φ_0 in Equation 4.1 generally contains a term describing the shape of the wave front as well as a time-dependent term that allows for variations in the phase of the laser, that is, its temporal coherence. For a fully coherent spherical wave, φ_0 may be expressed in terms of the curvature, $\varphi_0(\bar{r},t) = k\kappa(z)R^2$, and for a Gaussian beam wave the curvature is given in a paraxial approximation in terms of z and W_0 by

$$\kappa^{-1}(z) = 2z\left[1+\left(\frac{kW_0^2}{2z}\right)^2\right] \qquad (4.5)$$

This is a convenient mathematical model that will be used for purposes of illustration throughout the remainder of the book. It provides a realistic description of laser sources used in optical applications, but it does not account for the lobe structure often found in the emission from microwave and acoustic sources. However, these features will not greatly affect the nature of the fluctuation phenomena with which we are concerned here.

4.3   Moments of the Field Scattered by a Fixed Number of Small Particles

Figure 4.1 depicts the scattering of a beam wave from a collection of N discrete objects. The electromagnetic field impinging on a detector placed at position $\vec{r}$ is the sum of contributions scattered from each object. Each contribution can be represented by a phasor, and the resultant field is the sum of these phasors

$$E(\vec{r},t) = \sum_{j=1}^{N} a_j \exp(i\varphi_j) \tag{4.6}$$

Here

$$a_j = a(\vec{r}_j;\vec{r},t), \quad \varphi_j = \varphi(\vec{r}_j;\vec{r},t) \tag{4.7}$$

are the amplitude and phase of the field scattered from an individual object labelled by the subscript j. These will depend on the orientation and position of the object at a given time within the incident illumination defined by Equation 4.1. The sum of Equation 4.6 can also be represented as in Figure 4.2 by the addition of vectors in a plane: a two-dimensional walk of N steps. The magnitude and phase of E are given by the length and angle of a vector joining the origin to the final point in the walk. In square-law envelope

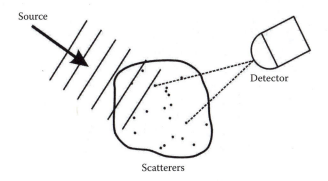

FIGURE 4.1
Scattering by a system of discrete objects.

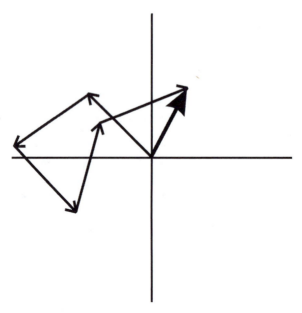

FIGURE 4.2
Sum of random phasors shown as a random walk on a plane.

detection, the output from the detector is proportional to the intensity, which is the square of the length of the vector or, equivalently, the squared modulus of the complex number E (see Section 1.8). The phases in this sum are determined by the path differences from the source to the detector via the individual scattering objects as well as the curvature and source terms present in Equation 4.1.

The simplest application of this model is in the description of scattering from a fixed number of identical, small (i.e., subwavelength), spherical particles that are randomly distributed over a uniformly illuminated volume much larger than the wavelength of the incident field. This last condition means that, except in near forward-scattering directions, the path differences that determine the phases are much greater than the wavelength, so that any value of phase is equally probable (this refers to the principal value of the phase: its value modulo 2π). Thus the probability of finding the steps in Figure 4.2 pointing in any given direction will be the same (the $\{\varphi_j\}$ will be uniformly distributed over 360°) and Equation 4.6 becomes an isotropic two-dimensional random walk of equal independent steps. With these simplifying assumptions, it is possible to calculate a number of useful statistical properties of the resultant field at a single space-time point. It can be seen immediately that the average value of E is zero because

$$\langle \exp(i\varphi_j) \rangle = 0 \tag{4.8}$$

It can also be seen that

$$\langle \exp(qi\varphi_j) \rangle = \delta_{q0} \tag{4.9}$$

That is, the averaged quantity in Equation 4.9 is only nonzero when $q = 0$. From this it follows that the only nonvanishing moments of E are those that can be expressed in terms of the intensity. This is because any moment of E is a sum of terms, each of which is a product of averages like Equation 4.9; the original average of products becoming equal to a product of averages because of the statistical independence of factors with different j values. The only way in which all of the factors in one or more of the terms can be nonzero (by having zero values of q) is if all the phases arise in pairs of opposite sign, which only occurs in the case of intensity moments. For example, the mean intensity can be written

$$\langle I \rangle = a^2 \sum_{j,k=1}^{N} \left\langle \exp\left[i(\varphi_j - \varphi_k)\right] \right\rangle = Na^2 \tag{4.10}$$

Here, the off-diagonal terms ($j \neq k$) of the sum vanish, but each diagonal term contributes. The second moment of the intensity is

$$\langle I^2 \rangle = a^4 \sum_{j,k,l,m=1}^{N} \left\langle \exp\left[i(\varphi_j - \varphi_k + \varphi_l - \varphi_m)\right] \right\rangle = 2N(N-1)a^4 + Na^4 \tag{4.11}$$

In this case the terms that survive fall into two categories: the pairs $\{j = k, l = m\}$ and $\{j = m, k = l\}$ that give $2N(N-1)$ terms, and the special case $j = k = l = m$ that gives N terms.

Suppose now that the scattering amplitudes in Equation 4.6, that is, the step lengths of the random walk, are not equal but are random quantities drawn from the same probability distribution, remaining statistically independent from each other and from the directions of the steps. This allows for more realistic situations: for example, scattering by different kinds of objects (such as particles of different sizes), or variations of intensity over the illuminated region. It is not difficult to generalize these results for this case. In particular, the normalized second moment takes the form

$$\frac{\langle I^2 \rangle}{\langle I \rangle^2} = 2\left(1 - \frac{1}{N}\right) + \frac{1}{N}\frac{\langle a^4 \rangle}{\langle a^2 \rangle^2} \tag{4.12}$$

It is evident that there are two kinds of contribution here: one dependent and one independent of the individual scattering amplitudes or step lengths in the random walk. Both contributions vary with the number of scatterers, but as the number of steps in the walk becomes large, the only term that survives is one that is *independent* of the properties of the scattering objects.

When there are a finite number of steps in the random walk, Equation 4.12 predicts that the deviation of the normalized second intensity moment from two is proportional to $1/N$ and will therefore typically be inversely proportional to the size of the illuminated region. It turns out that in most cases of practical interest, fluctuations in the magnitude of the scattering from each object will cause the final term in Equation 4.12 to be dominant so that the intensity fluctuations will *increase* as the illuminated region is decreased. This has important consequences, for example, for radar systems operating in the presence of background *clutter*, that is, unwanted reflections from the target environment. Thus while reducing the footprint or region illuminated by the radar will have the desirable effect of increasing the ratio of signal power to unwanted background clutter power, it may at the same time increase the degree of fluctuation of the clutter. In maritime radar limited by unwanted returns from waves on the sea surface, for example, this may reduce the discrimination of the system by reducing the difference in appearance between the target and clutter returns; effectively, reflections from individual waves begin to look like target returns [7].

4.4   The Probability Distribution of the Scattered Wave

The probability density of the complex amplitude of the field in Equation 4.6 can be most easily investigated through a calculation of the joint characteristic function of E. This is just the Fourier transform of the joint probability density of the real and imaginary parts

$$\left\langle \exp[i(u_R E_R + u_I E_I)] \right\rangle = \left\langle \exp\left[i \sum_{j=1}^{N} a_j (u_R \cos \varphi_j + u_I \sin \varphi_j)\right] \right\rangle$$

$$\hspace{6cm} (4.13)$$

$$= \left\langle \prod_{j=1}^{N} \exp[i a_j u \cos(\varphi_j - \theta_u)] \right\rangle$$

Here $u = \sqrt{u_x^2 + u_y^2}$, $\tan \theta_u = u_I / u_R$, and the averaging is carried out over realizations of the random amplitudes and phases of the scattering centers. If the contributions from different scatterers are statistically independent from one another, then the average of the product on the right can be expressed as the product of averages. Moreover, if it is assumed as before that the scattering amplitudes are statistically independent from the phases, then averaging each factor of the product over a uniform phase distribution leads to a product over zero-order Bessel functions of the first kind [8]

$$\langle \exp(i\bar{u}.\bar{E}) \rangle = \prod_{j=1}^{N} \langle J_0(ua_j) \rangle \tag{4.14}$$

Here we have replaced the complex representation used in Equation 4.13 by a more compact two-dimensional vector notation. Note that since Equation 4.14 depends only on the magnitude of u, it implies that the corresponding probability density function of E is circularly symmetric, or uniformly distributed in phase. This is to be expected, since it has been assumed that the contribution from each particle is uniformly distributed in phase. Thus, taking $E = A \exp(i\Phi)$ and assuming that the $\{a_j\}$ are drawn from a single statistical distribution, we may write $P_N(\Phi) = (2\pi)^{-1} (0 \leq \Phi \leq 2\pi)$ and, via the inverse Fourier transform of Equation 4.14,

$$P_N(A) = A \int_0^\infty u \, du J_0(uA) \langle J_0(ua) \rangle^N \tag{4.15}$$

The angled brackets now denote averaging over the probability density function of a. The probability density function of A^2 gives the intensity fluctuation distribution

$$P_N(I) = \frac{1}{2} \int_0^\infty u \, du J_0(u\sqrt{I}) \langle J_0(ua) \rangle^N \tag{4.16}$$

A result similar to Equation 4.16 was first obtained almost a century ago by Kluyver [1]. When the scattering amplitudes are identical constants, the integral can be expressed in terms of standard functions for $N < 4$. In this case, $N = 1$ gives a delta function on a as expected. Also, if a is constant, an analytical result for the case $N = 2$ can be obtained

$$P_2(I) = (2/\pi) \left[I(4a^2 - I) \right]^{-1/2} \quad 0 < I < 4a^2$$
$$= 0 \qquad\qquad\qquad \textit{otherwise} \tag{4.17}$$

This distribution is identical to that of the square of the amplitude of a randomly phased sine wave (Equation 1.13) calculated in Chapter 1. The simple U–shaped curve (Equation 4.17) is shown in Figure 4.3. The cutoff at $I = 4a^2$ expresses the fact that the maximum value of the sum (Equation 4.6) when $N = 2$ is $2a$ and occurs when the two contributions are in phase. Conversely the cutoff at $I = 0$ corresponds to the two contributions being exactly out of phase.

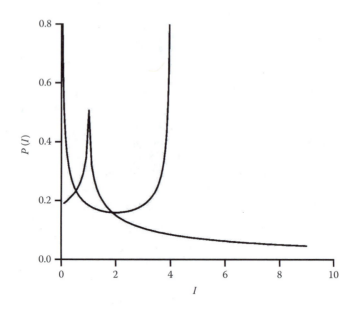

FIGURE 4.3
2-D random walk, probability densities for the cases of two (U-shaped curve) and three steps.

The integral on the right of Equation 4.16 can be expressed in terms of elliptic integrals when $N = 3$ (again a is constant) and is also plotted in Figure 4.3. Numerically computed distributions for $N = 4, 5$, and 6 are shown in Figure 4.4, using a logarithmic y axis. These were evaluated using the expansion quoted in Reference [9]. Evidently, fluctuations in the scattered radiation are sensitive to the number of scattering objects when that number is fairly small. This was recognized as long ago as 1940 and is recorded in a pilot training film that gave guidance on the way to estimate the number of unresolved aircraft from fluctuations in long wavelength radar returns [10]. A short theoretical paper on the topic was published in the 1960s [11].

It is clear from the log-linear plot (Figure 4.4) that as the number of scattering objects becomes large, the distributions approach a negative exponential shape. The central limit theorem can be applied separately to the rectangular components (real and imaginary parts) of Equation 4.6 in this limit, and predicts that they should tend to become statistically independent Gaussian variables (see Figure 4.5). As we have seen in Chapter 2 (Equation 2.20), this implies that the corresponding intensity has a negative exponential distribution. A simple approach to verify this result in the present context is to scale the scattering amplitudes a in Equation 4.16 with $\sqrt{N}$, so as to maintain the same total intensity, and expand the Bessel functions for large N [8]

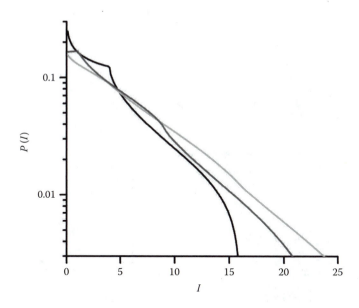

FIGURE 4.4
2-D random walk, probability densities for the cases of four (black), five (grey), and six (pale grey) steps.

$$\ln\left\langle J_0\left(\frac{ua}{\sqrt{N}}\right)\right\rangle^N = N\ln\left(1 - \frac{1}{N}\frac{u^2\left\langle a^2\right\rangle}{4} + \cdots\right) = -\frac{u^2\left\langle a^2\right\rangle}{4} + O\left(N^{-1}\right)$$

(4.18)

$$ie \quad \left\langle J_0\left(\frac{ua}{\sqrt{N}}\right)\right\rangle^N \approx \exp\left(-\frac{u^2\left\langle a^2\right\rangle}{4}\right)$$

When this result is substituted into the integrand in Equation 4.16, the integral may be performed and yields the result

$$P_\infty(I) = \frac{1}{\left\langle a^2\right\rangle}\exp\left(-\frac{I}{\left\langle a^2\right\rangle}\right)$$

(4.19)

The second normalized moment of this distribution is two, in agreement with the large N limit of Equation 4.12 and result (Equation 2.20) of Chapter 2.

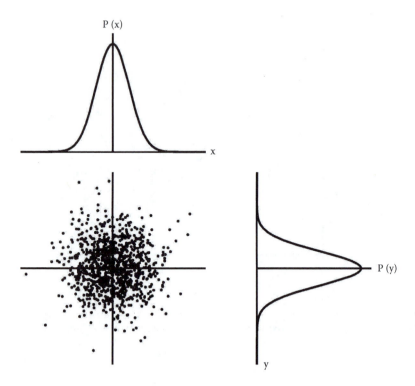

FIGURE 4.5
2-D random walk, end points of many realizations each of many steps.

4.5 Variations in Step Length: Illumination by a Gaussian Beam

Before going on to consider the correlation properties of the scattered field, we will look at extensions of the random walk model that involve variations in the step length (the present section) and step number (Section 4.6). In deriving the second moment result (Equation 4.12) and the general expression (Equation 4.15) for the probability density of the intensity scattered by N particles, we made no assumptions regarding the distribution of the individual particle scattering amplitudes a, save that they were statistically identical and independent from one another and from the associated phases. Simple analytical expressions for the higher moments can be obtained in this situation, and a number of models for the amplitude fluctuations have been used to evaluate these expressions [12]. If the particles are illuminated by a Gaussian beam, for example, according to Equation 4.3

$$a_j = b \exp\left(-\frac{R_j^2}{W^2}\right) \tag{4.20}$$

This assumes that the scattering amplitude is independent of scattering angle. The average signs in the earlier formulae indicate an average over all possible positions of the particles in the sample volume Ω. If this is limited to a cylinder of length L and radius R along the z axis, then [12]

$$\langle a^{2m} \rangle = \frac{b^{2m}}{\Omega} \int_0^L dz \int d^2R \exp\left(-\frac{2mR^2}{W^2}\right)$$

$$= \frac{\pi W^2 L}{\Omega} \frac{b^{2m}}{2m} = \frac{\Omega_{eff}}{\Omega} \frac{b^{2m}}{2m} \tag{4.21}$$

Here we have defined an *effective* scattering volume of length L and radius equal to the $1/e^2$ intensity radius of the beam. Result (Equation 4.21) can now be substituted into closed expressions for the moments of the intensity. In particular, for the second moment we obtain, from Equation 4.12

$$\frac{\langle I^2 \rangle}{\langle I \rangle^2} = 2\left(1 - \frac{1}{N}\right) + \frac{\Omega}{N\Omega_{eff}} \approx 2 + \frac{1}{\langle M \rangle} \tag{4.22}$$

In the final expression on the right-hand side, $\langle M \rangle$ is the *effective* average number of particles contributing to the scattered radiation. This may be quite small even though the total number of particles present is large and has been measured in sensitive light scattering experiments [13]. Thus, a very important feature of the random walk model is demonstrated: although the number of steps in the walk may be large, variations in step length may give rise to significantly enhanced fluctuations of the resultant by comparison with those of a Gaussian variable.

4.6 The Effect of Variations in Step Number

Another important model for the amplitudes $\{a_j\}$ arises from an idealization of the illuminated region, namely, that it has sharp boundaries rather than the "soft" Gaussian shape assumed above. This implies that the amplitudes either have a given value or are zero, according to whether the particles lie in the illuminated region or outside it. This is known as the *probability*

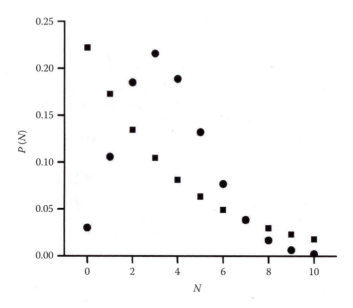

FIGURE 4.6
Scatterer number distributions: Poisson (circles) and negative binomial (squares) $\alpha = 1$.

aftereffect [14]. A completely equivalent description in this case is to characterize contributions to the scattered field by a *fluctuating number of steps* in the random walk. If it is assumed that the likelihood of the particles being in or out of the illuminated region is determined by Bernoulli trials (i.e., by coin tossing) then the number fluctuation distribution will be asymptotically Poisson (see Figure 4.6)

$$P(N) = \frac{\langle N \rangle^N}{N!} \exp(-\langle N \rangle) \tag{4.23}$$

Results such as Equation 4.11 and Equation 4.15 may be averaged over this distribution assuming that the fluctuations in N are statistically independent from those of the step lengths and directions. Equation 4.11 reduces after normalization to

$$\frac{\langle I^2 \rangle}{\langle I \rangle^2} = 2 + \frac{1}{\langle N \rangle} \frac{\langle a^4 \rangle}{\langle a^2 \rangle^2} \tag{4.24}$$

The intensity fluctuations are therefore always greater than the limit of negative exponential statistics for this kind of scattering system, but, as in the case of a fixed number of steps, the resulting fluctuations approach this limit as the mean number of contributions becomes large. This result is a consequence of the fact that the relative variance of a Poisson distribution decreases in inverse proportion to its mean

$$\frac{\text{var } N}{\langle N \rangle^2} = \frac{1}{\langle N \rangle} \tag{4.25}$$

so that the fluctuations vanish in the large mean limit. However, not all number fluctuation models behave in this way. For example, in the case of a negative binomial distribution, Figure 4.6, the variance remains finite as the mean becomes large

$$P(N) = \binom{N + \alpha - 1}{N} \frac{\left(\langle N \rangle / \alpha\right)^N}{\left(1 + \langle N \rangle / \alpha\right)^{N+\alpha}}$$

$$\frac{\text{var } N}{\langle N \rangle^2} = \frac{1}{\langle N \rangle} + \frac{1}{\alpha} \tag{4.26}$$

The persistence of the final term on the right-hand side of this result expresses the property of *clustering* embodied in the negative binomial model illustrated in Figure 4.7. This may occur in a population of seed particles due to the effect of shear in a turbulent flow or due to nonuniform mixing, for example. It may also be caused by modulation of the density of small wavelet scatterers by larger structures on a rough surface. It is interesting to consider the effect of negative binomial number fluctuations on the random walk. Averaging the right-hand side of Equation 4.15 over the distribution of Equation 4.26 leads to

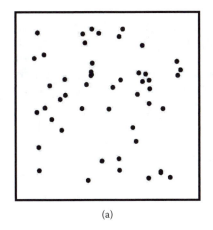
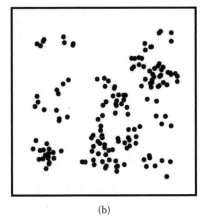

(a)   (b)

FIGURE 4.7
(a) randomly distributed scatterers, (b) clustered scattering centers.

$$P_{\langle N \rangle}(I) = \frac{1}{2} \int_0^\infty \frac{u\,du\,J_0(u\sqrt{I})}{\left(1 + \langle N \rangle [1 - \langle J_0(ua) \rangle]/\alpha\right)^\alpha} \tag{4.27}$$

If we now follow the procedure of Section 4.4, scale a with $\langle N \rangle^{-\frac{1}{2}}$ and expand the Bessel function for large *mean* number, similar to the procedure illustrated by Equation 4.18, we find in this limit [15]

$$P_\infty(I) = \frac{1}{2} \int_0^\infty \frac{u\,du\,J_0(u\sqrt{I})}{\left(1 + u^2 \langle a^2 \rangle / 4\alpha\right)^\alpha}$$

$$= \frac{2}{\Gamma(\alpha)} \left(\frac{\alpha}{\langle a \rangle^2}\right)^{\frac{\alpha+1}{2}} K_{\alpha-1}\left(2\sqrt{\alpha I / \langle a^2 \rangle}\right) \tag{4.28}$$

Thus, in the presence of this type of step-number fluctuation the asymptotic behavior of the random walk is not Gaussian, contrary to what one might expect from the central limit theorem. In particular, the intensity fluctuations are governed by the K-distribution (Equation 4.28), based on modified Bessel functions of the second kind. We have already encountered this distribution in the last chapter (Equation 3.23 and Equation 3.24, and Figure 3.7) where it was shown to characterize the fluctuations of the derivative of a gamma process and also those of a compound model in which the mean intensity of a complex Gaussian variable is modulated by a gamma distribution. The number fluctuation model of the present chapter is equivalent in the high mean limit to the latter compound process, since in the high density limit the discrete negative binomial distribution approaches the continuous gamma distribution. The class of K-distributions and the associated noise processes have many interesting analytical properties and have proved to be useful in a wide range of remote sensing applications where measurements are limited by non-Gaussian fluctuations [16]. A more comprehensive treatment of K-distributed noise is therefore reserved for a later chapter.

4.7   Intensity Correlation

Many optical diagnostic techniques such as photon correlation spectroscopy [4] and laser anemometry [5,17,18] are based on particle scattering. The measurement of fluid flow may use direct or coherent (homodyne, heterodyne) detection and may be based on having single or many seed particles in the illuminated volume. Evidently, if a large number of particles are involved, no information can be extracted from the probability density function of the intensity fluctuations, which is always of the form of Equation 4.19. Instead, information is derived from temporal or spatial correlation properties of the scattered field. The random walk model provides important insight into these properties. It is not difficult to show from Equation 4.6 that

$$\frac{\langle II' \rangle}{\langle I \rangle \langle I' \rangle} = \left(1 - \frac{1}{N}\right)\left(1 + \left|g^{(1)}\right|^2\right) + \frac{\langle a^2 a'^2 \rangle}{N\langle a^2 \rangle \langle a'^2 \rangle} \tag{4.29}$$

Here, the prime denotes a different point in space and/or time and

$$g^{(1)} = \frac{\langle EE'^* \rangle}{\sqrt{\langle I \rangle \langle I' \rangle}} \tag{4.30}$$

by analogy with the definition in Equation 2.23 and

$$\langle EE'^* \rangle = N\langle aa' \rangle \left\langle \exp\left[i(\varphi - \varphi')\right]\right\rangle \tag{4.31}$$

As N becomes large, the intensity correlation function (Equation 4.29) reduces to the Siegert relation (Equation 2.25), suggesting that E is a Gaussian *process* in this limit. It is not difficult to demonstrate that this is indeed the case by generalizing Equation 4.14 to include a label k representing different space-time points

$$\left\langle \exp\left(i\sum \vec{u}_k . \vec{E}_k\right)\right\rangle = \prod_{j,k} \langle J_0(u_k a_{j,k})\rangle \tag{4.32}$$

Following the procedure of Equation 4.18 shows that in the limit of large N

$$\left\langle \exp\left(i\sum \vec{u}_k . \vec{E}_k\right)\right\rangle = \exp\left(-\frac{1}{2}\left\langle\left(\sum_k u_k a_k\right)^2\right\rangle\right) \tag{4.33}$$

According to the discussion in Section 2.5, this is the expected result for a Gaussian process.

It is important to distinguish the physical mechanisms that lead to the different terms that occur in the result of Equation 4.29. The term in $|g^{(1)}|^2$ that remains in the Gaussian limit is a consequence of interference between the coherently added contributions to the scattered field. It is often referred to as the *speckle* contribution since, at visible wavelengths, a static Gaussian intensity pattern has a homogeneous speckled appearance. The spatial decay of this term, or speckle size, is just the square modulus of the Fourier transform of a projection of the scattering region (which can be regarded as a secondary source) on a plane normal to the line of sight from the scattering volume to the detector. This problem is addressed in many textbooks (see, for example, Reference [19]). It should be emphasized, however, that the existence of the interference term in Equation 4.29 was recognized long before the invention of the laser enabled the structure of the scattered intensity pattern to be seen by the naked eye. The review [9] contains references to early optics papers including one by Ramachandran [20]. This was published at about the same time as the radar work of Siegert in 1943 [21] and soon after it was recognized that microwave sea echo also displayed characteristic fluctuations that could be interpreted in the same way [22]. Perhaps the most famous application of the spatial behavior of the Gaussian term in Equation 4.29, however, was in the measurement of the angular diameter of stars. This was conventionally based on measurement of the size of the interference pattern embodied in the *first* order correlation function $g^{(1)}$ at two neighboring points on earth, assuming that the star can be regarded as a collection of random emitters. At radio frequencies, phase coherence over the required base line for such experiments was perfectly feasible, and direct measurement of $g^{(1)}$ could be carried out, but at optical frequencies it was impossible to achieve over the desired distances. The problem was solved by Hanbury, Brown and Twiss in the 1950s by exploiting the fact that the field correlation function is related to the intensity correlation through the Siegert relation [23].

Relative particle motion generally governs the temporal behavior of the interference term in Equation 4.29 and the functional form of this behavior depends on whether the particle velocities change over times that are smaller or greater than the reciprocal of the Doppler frequency shift that the light experiences in scattering from a moving particle. In the case of light scattering by a suspension of small particles diffusing under the influence of Brownian motion, for example, the former condition may hold, while in the case of seed particles used to delineate a fluid flow, the latter condition is more likely. To see how these conditions affect the correlation function, consider the simple case where it can be assumed that the particle velocity distribution is a stationary Gaussian process. From Equation 4.30 and Equation 4.31 and result in Equation 2.1, we obtain

$$\left\langle \exp\left\{i\left[\phi(0)-\phi(\tau)\right]\right\}\right\rangle = \left\langle \exp\left\{i\vec{K}.\int_0^\tau \vec{v}(t)dt\right\}\right\rangle =$$

$$\exp\left\{-\tfrac{1}{2}K^2\int_0^\tau dt\int_0^\tau dt'\left\langle v_x(t-t')v_x(0)\right\rangle\right\}$$

(4.34)

Here v_x is the velocity component along the scattering vector $K = 2k\sin\tfrac{1}{2}\theta$. If the particle velocities change much more rapidly than the corresponding reciprocal Doppler shift, then the velocity correlation function within the integral sign in Equation 4.34 will be sharply peaked about $t = t'$, and the formula reduces to

$$\left|g^{(1)}\right| = \exp\left(-K^2 D_T \tau\right) \tag{4.35}$$

where

$$D_T = \int_0^\infty dt \left\langle v_x(0)v_x(t)\right\rangle \tag{4.36}$$

is the translational diffusion coefficient. This result can be derived without the assumption of Gaussian statistics, provided that the distribution of particle velocities is isotropic [24]. It is the basis of the techniques of intensity fluctuation and photon correlation spectroscopy that are used at optical frequencies to determine the diffusion constant and hence the molecular weight of small particles in suspension, (Figure 4.8) [25]. An early review of these techniques is contained in References 4 and 5. There is now an enormous literature on light scattering measurements on complex particle systems, and it is beyond the scope of this book to give further coverage to this topic.

In the case of particle velocities that change slowly over the decay time of the field correlation function (Equation 4.34), v_x may be regarded as constant over the range of integration and we obtain

$$\left|g^{(1)}\right| = \exp\left(-\tfrac{1}{2}\left\langle v_x^2\right\rangle\tau^2\right) \tag{4.37}$$

Unlike Equation 4.35, this result is specific to the Gaussian velocity distribution model that is commonly used to characterize isotropic turbulence [26]. In fact, it will often be the case that particles move through the illuminated region with constant velocity. Then every contribution to the random

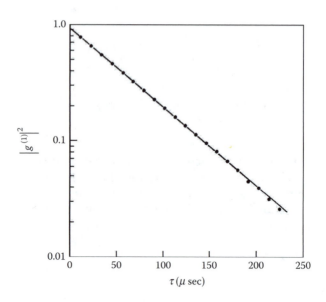

FIGURE 4.8
Intensity correlation function of light scattered by particles in suspension [25].

walk will carry the same Doppler shift within its phase. This can be measured
directly using coherent (heterodyne) detection as in Doppler radar or lidar
systems, for example. However, it will simply cancel in direct measurements
of *intensity* fluctuations. In other words, the modulus of Equation 4.30 will
be unity. In practice, the probability aftereffect will lead to amplitude fluctuations as the scatterers move into and out of the illuminated region, and
for uniformly moving particles the characteristic fluctuation time will be
equal to the time taken for particles to cross this region. A simple generalization of the procedure used in Section 4.6 gives the following expression
for radiation backscattered from particles moving across a Gaussian profile
beam with transverse velocity v [27] (see Figure 4.9)

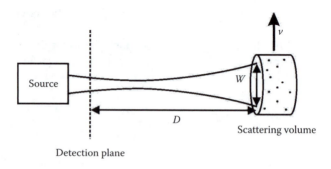

FIGURE 4.9
Backscattering from particles moving across an illuminating beam.

$$\left|g^{(1)}\right| = \exp\left(-\tau^2/2\tau_c^2\right)$$

$$\tau_c = W/v\sqrt{1+k^2\kappa^2W^4}$$

(4.38)

Note that a factor $\exp(i\kappa R_j^2)$ has been included in addition to Equation 4.20 to take account of the possibility of wave-front curvature. At the waist of the beam where the curvature vanishes, and also at very large propagation distances where it is small, the fluctuation time is just proportional to the residence time of the particles in the beam. The behavior in other regions is governed by the Gaussian beam propagation formulae (Equation 4.5). Result (Equation 4.38) provides one of the many optical techniques that can be used to measure fluid flow [18].

Some results have been derived in the Gaussian limit for the case where the correlation function (Equation 4.29) is measured at different points at different times. For example, the more general form of Equation 4.38 is [27]

$$\left|g^{(1)}\right| = \exp\left\{-\left(\tau-\tau_d\right)^2/2\tau_c^2\right\}\exp\left\{\left(\tau_d/\tau_c\right)^2 - \tfrac{1}{2}k^2V^2W^2\right\}$$

(4.39)

Here τ_c is given in Equation 4.38 and

$$\tau_d = k^2\kappa W^4\bar{v}.\bar{V}/v^2(1+k^2\kappa^2W^4)$$

$$\bar{V} = (\bar{R}-\bar{R}')/2D$$

(4.40)

where D is the distance from the scattering volume to the detection plane, and displacements are measured in this plane (see Figure 4.10). Therefore, $|g^{(1)}|$ is predicted to be a Gaussian function of τ of width τ_c centered at the delay time τ_d. Note that in the absence of curvature the correlation function is centered at zero delay. In this case an optical speckle pattern appears to *boil* due to particles moving into and out of the illuminated region. On the other hand, when there is wave-front curvature present, *bodily translation* of the speckle pattern due to geometrical leverage is observed as well as boiling.

A potentially important property of the result in Equation 4.39 is that it cannot be expressed as a product of separate space and time factors; that is, the conditions for cross-spectral purity are not satisfied [28]. This means that spatial averaging by the detector will change the apparent fluctuation time of the radiation. Conversely, time averaging will affect its transverse spatial coherence. Experimental measurement effects of this kind are discussed in Chapter 13.

The final term in Equation 4.29 quantifies the deviation from Gaussian statistics embodied in the Siegert factorization theorem (Equation 2.25) because the number of scattering centers are finite. Its structure reflects the

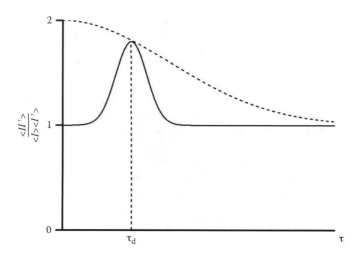

FIGURE 4.10
Intensity cross-correlation function obtained in the configuration Figure 4.9.

amplitude fluctuations of a single scattering center measured at two space-time points. Although this term augments the Gaussian fluctuations, in the case of a collection of identical small spherical particles, its spatial behavior is generally not very interesting as an additional source of information. This is because if the diffraction pattern from each scattering center is without spatial structure, the term will be spatially uniform. On the other hand, the temporal fluctuations will be determined by motion of the particles through the illuminated region and will always carry information about their dwell time. In the case of larger nonspherical objects, the non-Gaussian term will carry additional information about the shape of the scattering objects and their rotational characteristics as well as their movement through the illuminated volume.

The characteristic scale of the non-Gaussian contribution to the scattered intensity pattern will typically be much larger than the Gaussian speckle size that is determined by the size of the illuminated area. Thus, we expect that intensity patterns generated by only a few scattering objects will be characterized by two distinct scales, as shown in Figure 4.11. Similarly, the Gaussian speckle contribution to the intensity correlation function is dominated by phase fluctuations that will generally take place on a faster time scale than the amplitude fluctuations that govern the behavior of the final term in the result of Equation 4.29. The latter will reflect the actual time scales of the motion of individual scattering centers rather than the range of Doppler shifts present in the Gaussian term. Thus, we are led to expect the intensity correlation function for the non-Gaussian resultant of a finite random walk model for discrete scatterers to exhibit a delay time dependence that is also of the form shown in Figure 4.11; the rapid initial falloff being due to interference and the slower decay resulting from additional amplitude

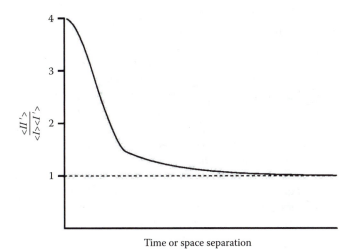

FIGURE 4.11
Form of correlation function expected when only a few scatterers are present.

fluctuations. Real data from a number of continuous scattering systems, ranging from stellar scintillation to microwave sea echo, has been found to behave qualitatively in this way [29–31]. This suggests that the random walk model provides at least a qualitative description of the fluctuation properties of radiation scattered by such systems. Studies of the more specific, but widely applicable, phase screen scattering model in the next chapter confirm this conjecture.

4.8   Partially Developed Speckle

In the foregoing sections, it has been assumed that the phase shifts introduced by each scattering object are uniformly distributed over 2π radians. It was argued in Section 4.2 that, except in near forward-scattering directions, this would be the case for scattering by a system of small particles that were distributed over a volume that was much larger than the wavelength of the incident radiation. In forward-scattering directions, small particles do not introduce large path differences into the incoming radiation and, indeed, calculations show that as the scattering angle is reduced, the uniformly distributed phase approximation breaks down in a way that depends on the number of scattering centers.

There is also another important scattering configuration in which the random phases in Equation 4.6 are not uniformly distributed, namely, when the scatterers are confined within a thin layer or deposited onto a rough surface. In near specular directions, the phase differences introduced by such systems may be quite small. This leads to the existence of a nonzero average *field* and an *unscattered* contribution to the mean intensity. Optical physicists refer to the intensity pattern in this case as *partially developed*.

It is possible to exploit the deviation from zero mean circular Gaussian statistics that arises if the random phases appearing in Equation 4.6 are not uniformly distributed to determine the probability density of the phases introduced by the scatterers or the number present within the illuminated area [32]. Some authors have envisaged using Equation 4.6 to model rough surface scattering itself, as opposed to the additional scattering from objects deposited onto a surface. In this picture, each contribution to the sum is interpreted as emanating from an independent element of the surface with a phase determined by the surface height. The lateral correlation length of the height variations governs the size of the independent elements and therefore the number present in the illuminated area. Thus both surface roughness and correlation length can be deduced from the measured statistics of scattered radiation. In practice, although experiments show qualitative agreement with the predicted dependence on height variance and number of surface elements illuminated, quantitative comparisons with computed predictions are generally poor. There were many contributions on this topic published in the 1970's stimulated by the early work of Beckmann and Spizzichino [33]. A useful list is to be found in Reference [34].

Surface scattering will be discussed further in later chapters where continuous scattering models that provide a more realistic description will be considered. However, it is important to establish the criteria under which the uniform phase approximation breaks down and its statistical consequences. Therefore, we shall extend some of the earlier analysis to determine the predictions of the discrete random walk model for scatterers that introduce path differences that are less than the wavelength of the incident radiation.

Assuming for purposes of illustration that the amplitudes in Equation 4.6 are unity and that the number of scatterers is fixed, we find by retaining the unmatched phase terms in the approach of Section 4.3 that

$$\langle E \rangle = N \langle \exp(i\varphi) \rangle \tag{4.41}$$

$$\langle I \rangle = N + N_2 \left| \langle \exp(i\varphi) \rangle \right|^2 \tag{4.42}$$

$$\langle I^2 \rangle = N + 2N_2 + 4(N_2 + N_3) \left| \langle \exp(i\varphi) \rangle \right|^2 + N_2 \left| \langle \exp(2i\varphi) \rangle \right|^2 + $$
$$+ N_4 \left| \langle \exp(i\varphi) \rangle \right|^4 + 2N_3 \operatorname{Re} \left[\langle \exp(i\varphi) \rangle^2 \langle \exp(-2i\varphi) \rangle \right] \tag{4.43}$$

Here $N_r = N(N-1)(N-2) \ldots (N-r+1)$. Expressions for higher intensity moments can be obtained but are lengthy and the reader is referred to the literature [34]. If we set the phase factors to zero at the outset, then Equation 4.42 and Equation 4.43 reduce to the results of Equation 4.10 and Equation 4.11. The formulae become more transparent when further averaged over the commonly encountered Poisson number fluctuations. Then, since $\langle N_r \rangle = \langle N \rangle^r = \bar{N}^r$, we obtain

$$\langle I \rangle = \bar{N} + \bar{N}^2 \left| \langle \exp(i\varphi) \rangle \right|^2 \tag{4.44}$$

$$\langle I^2 \rangle = \bar{N} + \bar{N}^2 \left[2 + 4 \left| \langle \exp(i\varphi) \rangle \right|^2 + \left| \langle \exp(2i\varphi) \rangle \right|^2 \right] +$$

$$+ 2\bar{N}^3 \left(2 \left| \langle \exp(i\varphi) \rangle \right|^2 + \mathrm{Re}\left[\langle \exp(i\varphi) \rangle^2 \langle \exp(-2i\varphi) \rangle \right] \right) + \tag{4.45}$$

$$+ \bar{N}^4 \left| \langle \exp(i\varphi) \rangle \right|^4$$

It is clear from Equation 4.44 and Equation 4.45 that even if the phase factors are small, for a sufficiently large number of scattering centers the phase-dependent terms cannot be neglected. In particular, in the large $\bar{N}$ limit, the final terms will dominate and the intensity fluctuations vanish asymptotically, that is, $\langle I^r \rangle \rightarrow \langle I \rangle^r$. The normalized second intensity moment given by these formulae when the phase is assumed to be a Gaussian variable, that is, $\langle \exp(in\varphi) \rangle = \exp(-n^2\sigma^2/2)$, is plotted in Figure 4.12 for various values of and $\bar{N}$.

The first contribution on the right-hand side of Equation 4.44 is identified as an incoherent or scattered contribution to the mean intensity, while the second term is an unscattered or coherent component. In Equation 4.45, the first and second terms can be recognized as the non-Gaussian and Gaussian contributions encountered in the uniformly distributed phase case treated in Section 4.6. A simple criterion for the validity of the assumption of uniform phase is that $\bar{N}|\langle \exp(i\varphi) \rangle|^2 \ll 1$. The breakdown of this inequality as a function of scattering angle has been investigated experimentally by scattering light from particles deposited on a silica glass plate [32].

We have seen that when the phases introduced by the scatterers are not uniformly distributed, the relative degree of intensity fluctuations becomes small in the large number limit and the scattered field is clearly not a zero-mean circular Gaussian process. On the other hand, the rectangular components (real and imaginary parts) of the field remain the sum of many independent contributions in this limit, and therefore tend to become Gaussian variables according to the central limit theorem. However, they will have nonvanishing mean values, variances that may be different, and they may also be correlated. It is not difficult to show that, in the presence of Poisson

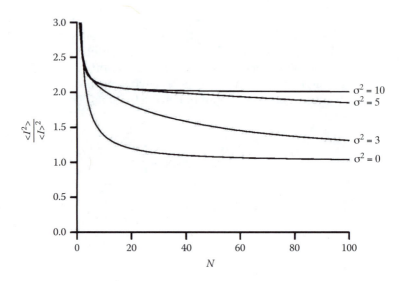

FIGURE 4.12
Contrast of partially developed speckle as a function of phase variance and scatterer number.

number fluctuations, the mean and variance of the probability distributions for the real and imaginary parts of the field (Equation 4.6) are given by the expressions (assuming unit scattering amplitudes for the purpose of illustration)

$$\langle E_R \rangle = \bar{N} \langle \cos \varphi \rangle, \quad \langle E_I \rangle = \bar{N} \langle \sin \varphi \rangle$$

$$\sigma_R^2 = \operatorname{var} E_R = \bar{N} \langle \cos^2 \varphi \rangle, \quad \sigma_I^2 = \operatorname{var} E_I = \bar{N} \langle \sin^2 \varphi \rangle \tag{4.46}$$

while the correlation coefficient is

$$c = \langle E_I E_R \rangle - \langle E_I \rangle \langle E_R \rangle = \bar{N} \langle \sin \varphi \cos \varphi \rangle \tag{4.47}$$

These results depend on the nature of the phase fluctuations and will fully characterize the Gaussian limit obtained when the mean number of scatterers is very large. In this limit it is always possible to construct linear combinations X, Y of the variables E_I and E_R that are uncorrelated Gaussian variables (see Chapter 2).

$$X = E_R \cos\theta + E_I \sin\theta, \quad Y = E_I \cos\theta - E_R \sin\theta$$

$$\tan 2\theta = 2c \Big/ \left(\sigma_R^2 - \sigma_I^2\right) \tag{4.48}$$

$$P(X,Y) = \frac{1}{2\pi\sigma_X\sigma_Y} \exp\left\{-\left[\frac{\left(X - \langle X\rangle\right)^2}{2\sigma_X^2} + \frac{\left(Y - \langle Y\rangle\right)^2}{2\sigma_Y^2}\right]\right\}$$

The intensity distribution may be expressed in terms of these new variables in the form of a sum of modified Bessel functions [35]

$$P(I) = \frac{f_1}{2\sigma_X\sigma_Y}\left\{I_0(f_2)I_0(f_3) + 2\sum_{n=1}^{\infty}I_n(f_2)I_{2n}(f_3)\cos 2n\alpha\right\} \tag{4.49}$$

where

$$f_1 = \exp\left\{-\left[\frac{1}{4}\left(\frac{1}{\sigma_X^2} + \frac{1}{\sigma_Y^2}\right)I + \frac{1}{2}\left(\frac{\langle X\rangle^2}{\sigma_X^2} + \frac{\langle Y\rangle^2}{\sigma_Y^2}\right)\right]\right\}$$

$$f_2 = -\frac{1}{4}I\left(\frac{1}{\sigma_X^2} - \frac{1}{\sigma_Y^2}\right), \quad f_3 = -\sqrt{I}\left(\frac{\langle X\rangle^2}{\sigma_X^2} + \frac{\langle Y\rangle^2}{\sigma_Y^2}\right)^{1/2} \tag{4.50}$$

$$\tan\alpha = \sigma_X^2 \big/ \sigma_Y^2$$

When the variances are equal, $f_2 = 0$, and Equation 4.49 reduces to the Rice distribution (Equation 3.5) the probability density of the phase derivative is given by Equation 3.8. The result is then the same as that obtained for the coherent sum of a constant signal and a zero-mean circular complex Gaussian process. It is therefore the same type of distribution that would be obtained if unscattered radiation were to be coherently mixed (interfered), either deliberately or inadvertently, with fully developed speckle noise — a situation frequently encountered in experimental measurements due to the presence of unwanted returns from background objects, and also in coherent (homodyne) detection systems. This case is also identical to the large step-number limit of the *biased* two-dimensional walk — a model that will be considered in more detail in a later chapter on K-distributed noise [36].

References

1. Kluyver, J.C. A Local Probability Problem. *Proc. Roy. Acad. Sci. Amsterdam,* 8 (1905): 341–50.
2. Pearson, K. A Mathematical Theory of Random Migration. *Draper's Company Research Memoirs,* Biometric Series III, no. 15 (1906).
3. Lord Rayleigh (Strutt, J.W.). On the Problem of Random Vibrations in One, Two, or Three Dimensions. *Phil. Mag.* 6 (1919): 321–47.
4. Cummins, H.Z., and E R Pike, eds. *Photon Correlation and Light Beating Spectroscopy.* New York: Plenum Press, 1974.
5. Cummins, H.Z., and E R Pike, Eeds. *Photon Correlation Spectroscopy and Velocimetry.* New York: Plenum Press, 1977.
6. Kogelnik, H., and T. Li. Laser Beams and Resonators. *Appl. Opt.* 5 (1966): 1550–67.
7. Ward, K.D., C.J. Baker, and S. Watts. Maritime Surveillance Radar, parts 1 and 2. *IEE Proc.* F 137 (1990): 51–69.
8. Abramowitz, M., and I.A. Stegun. *Handbook of Mathematical Functions.* New York: Dover, 1971.
9. Watts, T.R., K.I. Hopcraft, and T.R. Faulkner. Single Measurements on Probability Density Functions and Their Use in Non-Gaussian Light Scattering. *J. Phys. A: Math. Gen.* 29 (1996): 7501– 17.
10. Fisher, D., prod. *Weren't those great days,* a video potpourri of Wartime Radar Research at the Telecommunications Research Establishment (TRE), produced by Douglas Fisher, White House, Slough Road, Brantham Manningtree, Essex CO11 1N5, 1989.
11. Beckmann, P. Estimation of the Number of Unresolvable Targets from a Single Rradar Return. *Radio Science* 2 (1967): 955–60.
12. Pusey, P.N. Statistical Properties of Scattered Radiation. in Cummins, H.Z., and E.R. Pike, eds. *Photon Correlation Spectroscopy and Velocimetry.* New York: Plenum Press, 1977, 45–141.
13. Schaefer, D.W., and P.N. Pusey. Statistics of Non-Gaussian Scattered Light. *Phys. Rev. Letts.* 29 (1972): 843–45.
14. Chandrasekhar, S. Stochastic Problems in Physics and Astronomy. *Rev. Mod. Phys.* 15 (1943): 1–89.
15. Jakeman, E., and P.N. Pusey. Significance of *K*-Distributions in Scattering Experiments. *Phys. Rev. Lett.* 40 (1978): 546–48.
16. For a brief review see E. Jakeman, *K*-Distributed Noise. *J. Opt. A: Pure Appl. Opt.* 1 (1999): 784–89.
17. Schultz-Dubois, E.O., ed. *Photon Correlation Techniques in Fluid Mechanics.*Berlin: Springer-Verlag, 1983.
18. Durst, F., A. Melling, and J.H. Whitlaw. *Principles and Practice of Laser Doppler Anemometry.* London: Academic Press, 1976.
19. Dainty, J.C., ed. *Laser Speckle and Related Phenomena.* New York: Plenum, 1975.
20. Ramachandran, G.N. Fluctuations of Light Intensity in Coronae Formed by Diffraction. *Proc. Indian Acad. Sci.* A 18 (1943): 190–200.
21. Siegert, A.J.F. MIT *Rad. Lab. Rep.* No. 463 (1943).
22. Goldstein, H. The Origins of Echo Fluctuations. in *Propagation of Short Radio Waves.* Ed. D E Kerr. New York: Dover 1965, 527–50.

23. Hanbury Brown, R., and R.Q. Twiss. A Test of a New Type of Stellar Interferometer on Sirius. *Nature* 178 (1956): 1046–48.

24. Cummins, H.Z., and P N Pusey. Dynamics of Molecular Motion. in Cummins, H.Z., and E.R. Pike, eds. *Photon Correlation Spectroscopy and Velocimetry.* New York: Plenum Press, 1977, pp 164–99.

25. Pusey, P.N., D.E. Koppel, D.W. Schaefer, R.D. Camerini-Otero, and S.H. Koenig. Intensity Fluctuation Spectroscopy of Laser Light Scattered by Solutions of Spherical Viruses. *Biochemistry* 13 (1974): 952–60.

26. *Statistical Models and Turbulence.* Lecture Notes in Physics 12 New York: Springer, 1972.

27. Jakeman, E. The Effect of Wavefront Curvature on the Coherence Properties of Laser Light Scattered by Target Centres in Uniform Motion. *J. Phys. A: Math. Gen.* 8 L (1975): 23–28.

28. Goodman, J.W., *Statistical Optics.* New York: John Wiley, 1985, 187.

29. Goldstein, H. "Sea echo." In *Propagation of Short Radio Waves.* Ed. D E Kerr. New York: Dover, 1965, 481–527.

30. Jakeman, E., and P.N. Pusey. Non-Gaussian Fluctuations in Electromagnetic Radiation Scattered by a Random Phase Screen II: Application to Dynamic Scattering in a Liquid Crystal. *J. Phys. A: Math. Gen.* 8 (1975): 392–410.

31. Jakeman, E., E.R. Pike, and P.N. Pusey. Photon Correlation Study of Stellar Scintillation. *Nature* 263 (1976): 215–217.

32. Kazmierezak, M., T. Keyes, and T. Ohtsuki. Determination of the Number of Scatterers in a Finite Volume by the Statistical Analysis of Scattered Light Intensity. *Phys.Rev.* B 39 (1989): 1315–19.

33. Beckmann, P., and A. Spizzichino. *The Scattering of Electromagnetic Waves from Rough Surfaces.* London: Pergamon, 1963.

34. Uozumi, J., and T. Asakura. The First-Order Statistics of Partially Developed Non-Gaussian Speckle Patterns. *J. Optics* 12 (1981): 177–186.

35. Welford, W.T. First Order Statistics of Speckle Produced by Weak Scattering Media. *Opt. Quant. Electron.* 7 (1975): 413–16.

36. Jakeman, E., and R.J.A. Tough. Generalised *K*-Distribution: A Statistical Model for Weak Scattering. *J. Opt. Soc. Am. A* 4 (1987): 1764–72.

5

Scattering by Continuous Media: Phase Screen Models

5.1 Introduction

Just as the random walk provides a generic model for scattering by *discrete* objects, the random phase changing screen, or diffusing layer, exemplifies the main features of scattering by large-scale refractive index fluctuations in *continuous* media. By large-scale here it is meant that the inhomogeneities are much larger than the radiation wavelength. Figure 5.1 depicts the model in its simplest form: radiation impinges on a thin layer that introduces spatially random distortions or phase fluctuations into the incident wave front. Thus, the scattered field is given in terms of the incident one by

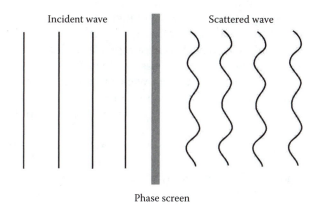

Phase screen

FIGURE 5.1
Scattering by a random phase changing screen.

$$E_s = E_0 \exp\left[i\varphi(\vec{r}, t) \right] \tag{5.1}$$

The field then propagates freely according to Maxwell's equations and exhibits a range of diffraction, interference, and geometrical optics effects that depend on the properties of the random phase variable, $\varphi(\vec{r}, t)$. In practice, of course, phase fluctuations will not be introduced into an incident wave at a single plane as in the idealization Equation 5.1 but over a region of finite thickness, albeit much thinner than the ensuing free propagation path. However, in many situations of interest the amplitude fluctuations that will be necessarily present after propagation through a refractive layer of finite thickness will be small so that Equation 5.1 will provide a good model for the scattering.

Originally conceived as a model for the fading of radio waves due to fluctuations in ionospheric layers [1], the random phase screen has been used to describe a very wide variety of phenomena. These include interplanetary scintillation of radio waves due to fluctuations in the solar wind, the twinkling of starlight at optical and radio frequencies, laser scintillation due to refractive index fluctuations in the atmosphere, light scattering by turbulent layers of liquid crystal, ultrasound scattering by body tissue, reverberation of acoustic waves propagating through the ocean, the scattering of electromagnetic and sound waves from rough surfaces, and even gravitational scintillation due to mass fluctuations in the universe. Reference to the literature on many of these topics will be made in succeeding chapters, but it is not feasible to attempt a full bibliography here. A list of early contributions to the field is to be found in the mini review by Zardecki [2].

This chapter outlines the main features of phase screen scattering geometries and discusses some asymptotic results that demonstrate how phase screen scattering relates to the random walk model. In following chapters, a more detailed examination will be carried out to show how the properties of the scattered radiation are dictated by assumptions made for the statistical and spectral properties of $\varphi(\vec{r}, t)$. It is important to appreciate the results of analytical investigations of phase screen scattering that have been carried out over the years. These give valuable insight into the phenomenology of the problem and thereby provide a guide to the choice of model that is suitable for a given application. However, the analysis is more difficult than was necessary to elucidate predictions of the random walk model. Properties of the scattered radiation will therefore often be illustrated using graphical presentations based on numerical calculations and the simulation techniques described in Chapter 14. The final section of the chapter outlines the relationship between the phase screen model and Kirchhoff theory of rough surface scattering. Rough surfaces occur widely in nature, science, and engineering, and provide many potential applications for the model.

5.2 Scattering Geometries

Figure 5.2 illustrates the *Fraunhofer region* or far field scattering geometry familiar from the last chapter where the illuminated area is much smaller than the first Fresnel zone of radius $\sqrt{\lambda z}$. As in the case of particle scattering, the parameters of this geometry that affect the fluctuation properties of the scattered wave are the number and character of the scattering centers (phase distortions) within the illuminated area, and additionally the angle of view. However, unlike the case of scattering by small particles that generally contributed to the field at the detector from the whole of the illuminated area, a detector placed *close behind* a phase screen may receive radiation from only part of the illuminated area. It is well known, for example, that the majority of the radiation contributing to the field at a point z along the axis of a propagating plane wave comes from the first Fresnel zone, with contributions from outside this region largely cancelling out. In the case of phase screen scattering, geometrical optics effects may also limit the field of view. This can be understood from Figure 5.3 where we show a facetted wave

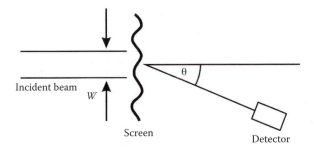

FIGURE 5.2
Observation in the Fraunhofer region.

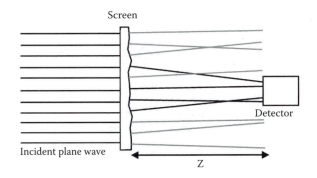

FIGURE 5.3
Observation in the Fresnel region.

front that could be generated by a phase grating consisting of adjoining prisms. By constructing rays normal to the facets, it is possible to delineate the trajectory of energy that contributes at a detector placed on axis some distance beyond the scattering layer. Since the tilts on the wave front are finite and most likely near zero, it is clear that most of the energy received by the detector will emanate from a narrow region near the axis, no matter how large an area is illuminated. Moreover, as in the case of a planar source, the closer the detector is placed to the screen, the smaller will be the region from which the radiation is collected.

In the case of phase screen scattering, therefore, when the illuminated area is much larger than the first Fresnel zone (in the case of the Gaussian beam profile Equation 4.3, this would be $W^2 \gg \lambda z$), the fluctuation properties of the scattered wave are parameterized by the propagation distance and are independent of the illuminated area and angle of view. This geometry, in which a phase screen of effectively infinite extent is illuminated by a plane wave, is the configuration relevant to ionospheric scattering where the model first found application [1]. We shall term the region of scattering beyond the screen, $W^2/\lambda \gg z \gg \lambda$, where the scattering is independent of the illuminated area, the *Fresnel region*. In this regime, the wave field beyond the screen obtained from the Helmoltz solution of the scalar wave equation (see Equation 5.22) in a paraxial approximation is [3]

$$E(\bar{r},t) = E_0 \left(\frac{ik\kappa}{\pi} \right) \int d^2 R' \exp[i\varphi(R',t)] \exp\left[-ik\kappa \left(R - R' \right)^2 \right] \qquad (5.2)$$

$$\kappa = \frac{1}{2}\left(\frac{1}{z} + \frac{1}{\sigma} \right) = \frac{1}{2z_{eff}}$$

Here, $\bar{r} \equiv (\bar{R}, z)$ where z is the propagation distance (Figure 5.3) and $\bar{R}$ is a transverse vector in the detection plane. Apart from a constant phase factor, Equation (5.2) is effectively Huyghens principle with the approximation

$$k\left|\bar{r} - \bar{R}'\right| = k\sqrt{z^2 + R^2 + R'^2 - 2\bar{R}.\bar{R}'} \quad \approx kz + k\left(\bar{R} - \bar{R}'\right)^2 \Big/ 2z \qquad (5.3)$$

valid sufficiently close to the axis of propagation. The parameter κ is the curvature of the propagating wave front and is given in terms of the radius σ of an incident spherical wave. In some situations this curvature factor can be changed, and so the *effective* propagation distance, z_{eff}, can be controlled. This may enable the full range of propagation regimes to be explored, for example, by using a lens in the case of laboratory optical frequency experiments.

The Fresnel region configuration is mathematically more tractable than the Fraunhofer geometry and has been the subject of a great deal of theoretical investigation since the 1950s. The earliest theoretical analysis of *fluctuation* properties in the Fresnel region of a random phase screen is attributed to Mercier in 1962 [3]. This was followed by a plethora of papers, including a notable contribution by Salpeter [4], spanning the next two decades, and the problem continues to attract interest because it constitutes the basic element in modelling wave propagation through extended inhomogeneous media (see Chapter 9). The first analysis of fluctuations in waves scattered by a deep random phase screen, $\langle \varphi^2 \rangle \gg 1$, into the Fraunhofer region did not appear in the Western literature until the early 1970s [5], and this geometry has received less attention, although it is relevant both to beam wave propagation and to many surface scattering configurations.

5.3   Time-Dependent Scattering

Although a notional time evolution is attributed to the phase fluctuations depicted in Equation 5.2, the random phase screen is most frequently used to model situations in which the spatial structure of the screen does not evolve with time but moves rigidly across the field of view with uniform velocity. This is often a good approximation in the case of scattering by turbulent media where it is known as the *Taylor hypothesis* of "frozen flow" [6]. It is also a valid assumption in the case of scattering by moving rigid rough surfaces that provide a large and important range of applications for the phase screen model. In such situations, $\varphi(\bar{r}, t + \tau) = \varphi(\bar{r} + \bar{v}\tau, t)$, where $\bar{v}$ is the transverse velocity, and the *temporal* fluctuations in the scattered field are thus dictated by the *spatial* structure of the screen. The linear transformation $\bar{R}' \rightarrow \bar{R}' - \bar{v}_T \tau$ in the integral on the right-hand side of Equation 5.2, where $\bar{v}_T$ is the transverse component of velocity, confirms that in the Fresnel geometry

$$E(\bar{r}, t + \tau) = E(\bar{r} + \bar{v}_T \tau, t) \tag{5.4}$$

This implies that a simple translation of the scattered radiation pattern across the propagation path occurs. The situation is different in the Fraunhofer region (Section 5.5) where the pattern may evolve as well as translate even though the phase screen itself executes only a simple translational motion. This is because individual phase inhomogeneities or "scattering centers" are constantly moving into and out of the illuminated region like particles in the discrete scattering model described previously (Section 4.7).

There appears to be relatively little analysis in the literature of scattering by a phase screen that *evolves* during the observation time. This kind of model was relevant to early light scattering measurements on electrohydrodynamic turbulence in nematic liquid crystals, and some results on this system have been published [5,7]. However, temporal evolution is also important in scattering by mobile rough surfaces such as the sea. It will be shown in following sections that in some situations phase screen scattering leads to Gaussian speckle. In such circumstances, the evolution of the scattering medium can be incorporated by adapting the discrete models discussed in the last chapter, for example, by including an element of diffusion as well as translation in the scatterer motion.

5.4   Scattering into the Fresnel Region

A number of simple results can be derived from Equation 5.2 for the case of phase fluctuations that are spatially and temporally stationary. For example, the field correlation function is given by

$$\left\langle E(\bar{r},t)E(\bar{r}+\bar{\Delta},t+\tau)\right\rangle =$$

$$|E_0|^2 \left(\frac{k\kappa}{\pi}\right)^2 \iint d^2R_1 d^2R_2 K\left(\bar{R}_1-\bar{R}_2,\tau\right) \tag{5.5}$$

$$\exp\left[ik\kappa\left(\bar{R}_1-\bar{R}_2-\bar{\Delta}\right)\left(2\bar{R}+\bar{\Delta}-\bar{R}_1-\bar{R}_2\right)\right]$$

Here

$$K\left(\bar{R}_1-\bar{R}_2,\tau\right)=\left\langle\exp\left\{i\left[\varphi\left(\bar{R}_1,t\right)-\varphi\left(\bar{R}_2,t+\tau\right)\right]\right\}\right\rangle \tag{5.6}$$

If φ is spatially stationary, this depends only on the difference coordinate. The displacement $\bar{\Delta}$ is measured transverse to the direction of propagation. Integrating over the sum coordinate yields a delta function on the difference coordinate whence the mean intensity $\langle I\rangle = |E_0|^2$ is independent of time and position. This simply confirms that on average the energy of the incident wave is conserved across any plane normal to the propagation direction. The normalized field autocorrelation function is given by

$$g^{(1)}(\bar{\Delta},\tau)=K(\bar{\Delta},\tau)=\left\langle\exp\left\{i\left[\varphi(0,0)-\varphi\left(\bar{\Delta},\tau\right)\right]\right\}\right\rangle \tag{5.7}$$

Therefore the form of the transverse field correlation function in the Fresnel region behind a stationary random phase screen is independent of propagation distance, being identical to that introduced at the screen.

An analytic result for the intensity correlation function can also be obtained from Equation 5.2 in certain circumstances. Provided that the phase fluctuations are spatially and temporally stationary, the full eightfold integral can be reduced to a fourfold integral by a process similar to that used to evaluate Equation 5.5

$$\frac{\langle I(0,0)I(\bar{\Delta},\tau)\rangle}{\langle I\rangle^2} =$$

$$\left(\frac{k\kappa}{\pi}\right)^2 \iint d^2r d^2r' \left\langle \exp\left\{i\left[\varphi(\bar{r}',0) - \varphi(\bar{r}+\bar{r}',0) + \varphi(\bar{r},\tau) - \varphi(0,\tau)\right]\right\}\right\rangle \quad (5.8)$$

$$\exp\left[-2ik\kappa\bar{r}.(\bar{r}'+\bar{\Delta})\right]$$

Now make the transformations $\bar{r} \to \bar{r}/\sqrt{\kappa}$, $\bar{r}'+\bar{\Delta} \to \bar{r}'/\sqrt{\kappa}$ and consider the behavior of the averaged quantity in the integrand as the effective propagation distance becomes large, that is, as $\kappa \to 0$. If the phase fluctuations are *independent* from one another at sufficiently large spatial separations, then this quantity has the asymptotic behavior [8]

$$\langle\cdots\rangle \to \delta_{r,0} + \delta_{r',0} \left|\left\langle\exp\left\{i\left[\varphi(0,0)-\varphi(\bar{\Delta},\tau)\right]\right\}\right\rangle\right|^2 + (1-\delta_{r,0}-\delta_{r',0})\left|\exp\left(i\varphi\right)\right|^4$$

$$(5.9)$$

Although the Kronecker delta function terms contribute to the integral (Equation 5.8) only along infinitesimally thin lines, Dirac delta functions that arise during the integration ensure that they make a finite contribution to the final result. Using Equation 5.7 and Equation 5.9, the result (Equation 5.8) may be expressed in the following form

$$\lim_{\kappa\to0} g^{(2)}(\bar{\Delta},\tau) = 1 + \left|g^{(1)}(\bar{\Delta},\tau)\right|^2 - \left|\langle\exp(i\varphi)\rangle\right|^4 \quad (5.10)$$

Note that the final term is negligible in the case of a *deep* random phase screen that introduces path differences much larger than the radiation wavelength into the incident wave front. In this case (Equation 5.10) is identical to the Siegert relation (Equation 2.25) that was encountered as a factorization property of circular complex Gaussian processes.

Equation 5.10 is a generalization of the result obtained by Mercier [3] for the case where the phase fluctuations themselves constitute a joint Gaussian

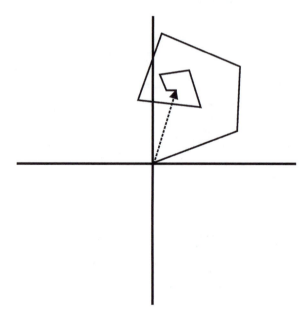

FIGURE 5.4
Random Cornu spiral, weak scattering into the Fresnel region.

process. Mercier was also able to show, by combinatorial analysis for this model, that at large propagation distances the scattered field amplitude fluctuations are governed by the Rice distribution discussed in Chapter 3 (Equation 3.4 and Equation 3.5). This appears to be a particular manifestation of a more general result that holds far from the scattering screen in the Fresnel limit *provided that the phase fluctuations are independent at sufficiently large separations.* A simple intuitive argument can be used to show this with the help of Figure 5.4. As already mentioned, sufficiently far from the screen, the scattered field is composed of many independent contributions of varying phase and amplitude. In the Fresnel limit these contributions acquire a systematic phase factor from the final exponential term on the right-hand side of Equation 5.2. Thus, when added together they produce a 'Cornu' spiral in the complex plane, as shown in Figure 5.4. When the screen introduces large random phase shifts, this systematic factor will be unimportant by comparison with the large random changes of direction associated with each contribution. The scattered field is then the result of a two-dimensional random walk, just as in the case of scattering by many discrete particles; it will be Gaussian by virtue of the central limit theorem, and the associated time-dependent field will be a zero-mean circular complex Gaussian process. When the phase screen is weak, the systematic phase factor imparts a spiral structure to the vector addition as illustrated in the figure. The mean value of the resultant complex amplitude is nonzero in this case. Since there are many contributions and many turns of the spiral, however, the real and imaginary part of this vector will again be Gaussian distributed and uncorrelated with equal variances. Hence the amplitude of the scattered field will be Rice distributed as found by Mercier [3]. In particular [8]

$$\lim_{\kappa \to 0} P(I) = \frac{\exp[-(I + |\langle E \rangle|^2)/2 \, \text{var}(\text{Re } E)]}{2 \, \text{var}(\text{Re } E)} I_0 \left(\frac{|\langle E \rangle| \sqrt{I}}{\text{var}(\text{Re } E)} \right) \tag{5.11}$$

Using the fact that the variances of the real and imaginary parts of E are equal, it may be shown that

$$\text{var}(\text{Re } E) = \tfrac{1}{2}(|E_0|^2 - |\langle E \rangle|^2) \tag{5.12}$$

while from Equation 5.2

$$\langle E \rangle = E_0 \langle \exp(i\varphi) \rangle \tag{5.13}$$

Note that Equation 5.11 applies at all points across the wave field sufficiently far from the phase screen, and that the associated time-dependent field will be a Rice process (Section 3.3).

A particularly important measure encountered in the phase screen scattering literature is the *scintillation index, S*. This is a measure of the degree of fluctuation identical to the average contrast, C, used to characterize an optical pattern. These quantities are related to the normalized second moment of the intensity fluctuation distribution through the definition

$$S^2 = C^2 = \frac{\langle I^2 \rangle}{\langle I \rangle^2} - 1 \tag{5.14}$$

From Equation 5.11 to Equation 5.13, far beyond the screen,

$$\lim_{\kappa \to 0} S^2 = 1 - |\langle \exp(i\varphi) \rangle|^4 \tag{5.15}$$

Equation 5.10 reduces to this result upon setting $\Delta, \tau = 0$. As expected, the contrast of the pattern is unity when the phase fluctuations are so large that the final term on the right-hand side of Equation 5.15 vanishes. The intensity distribution is then negative exponential with second normalized moment of two. As mentioned in Chapter 4, the intensity variations are often described in the optical literature in this case as constituting a fully developed speckle pattern, while if there exists a residual constant or unscattered component leading to Rice statistics, the speckle pattern is termed partially developed.

It has been demonstrated here that the statistics of the fluctuations in a wave field that has been scattered into the Fresnel region by a random phase screen evolve from a situation near the screen, where there are only phase fluctuations, to Gaussian at sufficiently large distances. Here many different elements of the screen contribute independently, the central limit theorem can be applied as in the random walk model, and the properties of the scattered field are qualitatively the same as those of radiation scattered by a large collection of particles distributed over the plane of the phase screen. The analog with particle scattering is particularly close if the phase fluctuations introduced by the screen are equivalent to path differences that are much larger than the radiation wavelength, since this then introduces the third "volume" dimension usually encountered in particle scattering configurations. It should be noted, however, that in the Fresnel region geometry, the coherence properties (Equation 5.7 and Equation 5.10) continue to express spatial characteristics of individual elements of the scattering screen. This may be contrasted with particle scattering into the Fraunhofer region where the spatial coherence properties are dictated only by the size of the illuminated volume, as described in Section 4.7. It is also important to note that the Gaussian limit is based on the assumption that at large enough propagation distances, many *independent* contributions reach the detector. It will be shown in Chapter 7 that in the case of ideal fractal phase screens characterized by many length scales, this may not be the case.

5.5   Fraunhofer Scattering

In view of this discussion, it comes as no surprise to find that in the more familiar Fraunhofer or far field geometry, where the illuminated area is much smaller than the first Fresnel zone, phase screens will also generate Gaussian speckle patterns, provided the illuminated area contains many independent scattering elements. This geometry is illustrated in Figure 5.2; whereas in the Fresnel region the statistics of the scattered field are a function of propagation distance, in the Fraunhofer region they vary with illuminated area and angle of view [8]

$$E(\bar{r},t) = \frac{ikE_0}{2\pi r} \int d^2R'\, A(\bar{R}') \exp\left\{ i\left[\varphi(\bar{R}',t) - \bar{k}.\bar{R}' \sin\theta \right] \right\} \qquad (5.16)$$

This result is again derived from the Helmoltz solution (5.22) of the scalar wave equation and is basically Huyghens principle, but now making the far field approximation

$$k\left|\bar{r} - \bar{R}'\right| = k\sqrt{r^2 + R'^2 - 2\bar{r}.\bar{R}'} \quad \approx kr - \bar{k}'.\bar{R}'\sin\theta \qquad (5.17)$$

Strictly speaking, E_0 is multiplied by an angle-dependent factor that is determined by the exact nature of the boundary condition imposed in the plane of the screen. For example, if it is assumed that the normal gradient of the phase is zero there, then this factor is simply $\frac{1}{2}(1 + \cos\theta)$. However, as this will not affect the normalized fluctuation statistics that is the main concern here, it will not be explicitly included in the formalism.

It will generally be assumed that $A(\bar{R})$ is a real aperture function, although it may be appropriate in some applications to include both aperture and wave-front curvature effects in this term. In the case of laser beam scattering, it may also be appropriate to use the Gaussian intensity profile (Equation 4.3). This model will often be used in the analyses presented in following chapters since it generally leads to a useful simplification of the calculations.

It is not difficult to show from Equation 5.16 that for a spatially and temporally stationary phase screen, provided that the decay length of the kernel (Equation 5.6) is small compared to the aperture size, the angular correlation function of the field may be expressed in the form

$$\left\langle E(\theta,t)E^*(\theta',t)\right\rangle = F\tilde{I}\left(\tfrac{1}{2}k\left[\sin\theta - \sin\theta'\right]\right)\tilde{K}\left(\tfrac{1}{2}k\left[\sin\theta + \sin\theta'\right]\right) \qquad (5.18)$$

Here

$$\tilde{I}(\bar{q}) = \int d\bar{R}A^2(\bar{R})\exp(i\bar{q}.\bar{R}), \quad \tilde{K}(\bar{q}) = \int d\bar{R}K(\bar{R},0)\exp(i\bar{q}.\bar{R}) \qquad (5.19)$$

are Fourier transforms of the aperture intensity function and the kernel (Equation 5.6), respectively. The value of F is only weakly dependent on angle in near forward scattering directions. The behavior of the mean intensity with angle ($\theta = \theta'$) at moderate angles is thus governed by the Fourier transform of the coherence function at the screen with argument $k\sin\theta$. Unlike the Fresnel region result (Equation 5.7), the characteristic spatial scale implicit in Equation 5.18 is determined by the source size through the aperture function.

It is important to note that the restriction on the characteristic decay length of the kernel needed to obtain the result in Equation 5.18 is equivalent to requiring the phase variation across the aperture to be much greater than the radiation wavelength. The restriction will therefore be satisfied even when the aperture is significantly smaller than individual phase inhomogeneities, provided the phase screen is deep (i.e., introduces path differences much greater than the radiation wavelength). However, if this restriction is not satisfied then aperture diffraction will begin to dominate the distribution of scattered intensity. Indeed, it is clear from Equation 5.16 that as the aperture is reduced towards a situation where the phase does not change significantly within it, then no amplitude fluctuations will be generated in the scattered field. On the other hand, when the illuminated area contains many independent phase inhomogeneities, the random walk picture will be applicable and the far field will be a zero-mean circular complex Gaussian process in the deep phase screen limit. Results for intensity correlations consistent with this conjecture are obtained when the aperture is large by comparison with the phase inhomogeneities by using approximations of the type

$$
\left\langle \exp\left\{ i\left[\varphi(\vec{R}_1, t) - \varphi(\vec{R}_2, t) + \varphi(\vec{R}_3, t) - \varphi(\vec{R}_4, t) \right] \right\} \right\rangle \approx
$$
$$
\delta^2(\vec{R}_1 - \vec{R}_2)\delta^2(\vec{R}_3 - \vec{R}_4) + \delta^2(\vec{R}_1 - \vec{R}_4)\delta^2(\vec{R}_2 - \vec{R}_3)
$$

(5.20)

This is just a continuum analog of the property used in the last chapter to obtain the second intensity moment (Equation 4.11) from Equation 4.6, and in the large aperture limit generalizations can be used to good effect in the calculation of higher moments and coherence properties. For example, it is possible to establish the Siegert factorization theorem (Equation 2.25) from Equation 5.16 using Equation 5.20

$$
\left\langle I(\theta, t)I(\theta', t) \right\rangle = \left\langle I(\theta, t) \right\rangle\left\langle I(\theta', t) \right\rangle + \left| \left\langle E(\theta, t)E^\bullet(\theta^1, t) \right\rangle \right|^2
$$

(5.21)

When there are many independent *weak* phase fluctuations within the illuminated area, it is possible to derive results for the partially developed speckle case by analogy with the approach of Section 4.8. Calculations for an application of phase screen scattering to imaging are described in Reference [9].

Even if the illuminated area is large and contains many independent scattering inhomogeneities, it is possible for significant deviations from Gaussian statistics to be observed in the Fraunhofer geometry. This may be understood from Figure 5.5 that shows the screen model of Figure 5.3 observed in the far field. It is self-evident that fewer of the wave-front elements will contribute at high viewing angles because the probability of finding appropriately oriented facets on the scattered wave front will be reduced. Thus the degree of fluctuation of the amplitude of the scattered field will generally increase with angle. In the Fraunhofer region, then, it is necessary to identify the *effective* number of scattering centers as a function of viewing angle.

A final point to note is that by inspection of Equation 5.16, it is evident that in the Fraunhofer geometry, rigid transverse motion of the screen does not lead to a simple translation of the intensity pattern as in the Fresnel region, Equation 5.4. Rather, the field evolves with time. As mentioned in Section 4.7, in the Gaussian speckle limit this has led to the evocative

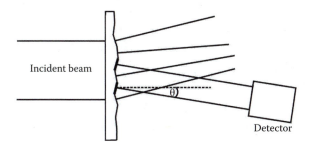

FIGURE 5.5
Observation off axis in the far field.

description of the pattern as appearing to boil, a consequence of scattering centers moving into and out of the illuminated region. In general, both translation and evolution of the intensity pattern will be observed according to the relative importance of curvature and finite aperture effects.

5.6   Scattering in Non-Gaussian Regimes

It has been demonstrated here that scattering by a random phase changing screen can lead to Gaussian statistics in the same way as particle scattering. In the Fresnel region, at the screen, there are no amplitude fluctuations, but these are generated during propagation of the scattered wave and become Gaussian at large distances when many independent elements of the screen contribute to the scattered field. In the Fraunhofer region, there are no fluctuations if the aperture is so small that within it the phase changes by much less than a wavelength, but fluctuations grow as the aperture is increased and again become Gaussian when, for a given viewing angle, the number of effective independent inhomogeneities is large.

For both geometries, the question naturally arises as to what the statistics of the scattered field are between the two well-understood limiting situations. A simple measure of this is the behavior of the scintillation index with propagation distance (i.e., *effective* illuminated area) in the Fresnel region or with the actual illuminated area in the Fraunhofer region. Figure 5.6 illustrates the trend generally observed and shows a marked peak in the degree of fluctuation before asymptotic convergence to the Gaussian value of unity. However, the detailed behavior in the intermediate regions is a strong function of the statistical and correlation properties of the screen, particularly in the case of deep phase screens. These can generate very large intensity fluctuations that may present a severe limit on system performance, or equally provide useful information about the phase screen itself. It is the statistical properties of waves scattered by deep random phase screens in these intermediate regimes, where the effective number of contributing

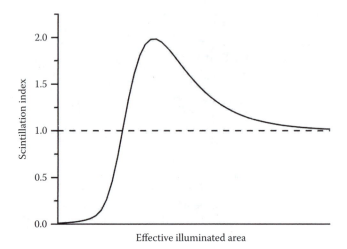

FIGURE 5.6
The scintillation index as a function of effective illuminated area.

scattering centers is small, which is of most interest. The following three chapters will therefore be devoted to the task of elucidating the relationship between the statistical models adopted for the phase and those predicted for the scattered wave.

5.7   Surface Scattering

One important group of scattering objects that has been modelled by phase screens is random rough surfaces. In this section, a brief outline of the underlying theory will be given and in particular the relationship between the parameters of the two kinds of objects will be established. The presentation is not intended to be a rigorous or comprehensive review of surface scattering, but designed to give an elementary understanding of how the phase screen model fits into the general picture of this large research area. The approach taken in the book by Ogilvy [10] based on the early work of Beckmann and Spizzichino [11] will be followed here. The scattering geometry is illustrated in Figure 5.7.

In this approach, the Helmoltz solution of the scalar wave equation for the field scattered by a surface S is given in terms of the incident and total fields by

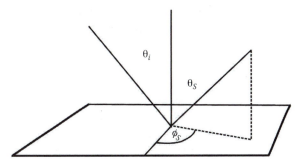

FIGURE 5.7
Surface scattering geometry.

$$E(\vec{r}) = E_{tot}(\vec{r}) - E_0(\vec{r})$$

$$= \int_S dS \left[E_{tot}(\vec{r}') \frac{\partial G(\vec{r},\vec{r}')}{\partial n'} - G(\vec{r},\vec{r}') \frac{\partial E_{tot}(\vec{r}')}{\partial n'} \right] \tag{5.22}$$

Here, the integral is carried out over the local values and normal gradients of E_{tot} and G at the surface denoted by the primed coordinates. The Green's function is given by

$$G(\vec{r},\vec{r}') = \frac{\exp\left(ik|\vec{r} - \vec{r}'| \right)}{4\pi|\vec{r} - \vec{r}'|} \tag{5.23}$$

The only unknown in this formula is the surface field. In the case of reflection of *plane* waves from a *planar* surface

$$E_{tot}(\vec{r}') = \left[1 + \Re \right] E_0(\vec{r}') \tag{5.24}$$

The reflection coefficient, $\Re$, in this formula depends on the angle of incidence. If the incident field is given by

$$E_0(\vec{r}) = \exp\left(i\vec{k}_0.\vec{r} \right) \tag{5.25}$$

then

$$\left(\frac{\partial E_{tot}(\vec{r}')}{\partial n'} \right) = i\left[1 - \Re \right](\vec{k}_0.\hat{n}')E_0(\vec{r}') \tag{5.26}$$

In the far field where Equation 5.17 is satisfied, the normal derivative of the Green's function (Equation 5.23) at the surface can be written in terms of the scattered wave vector $\bar{k}_s$ as follows

$$\frac{\partial G(\bar{r},\bar{r}')}{\partial \bar{r}'} = \frac{-i\exp(ik_0 r)}{4\pi r}(\hat{n}'.\bar{k}_s)\exp\left(-i\bar{k}_s.\bar{r}'\right) \tag{5.27}$$

The *tangent plane* approximation [10] is now made by applying expressions of Equation 5.24 and Equation 5.26 for plane surfaces *locally* within the integral Equation 5.23. Thus the scattered field (Equation 5.22) may be reduced to the form

$$E(\bar{r}) = \frac{i\exp(ik_0 r)}{4\pi r} \int_S dS\left[(\Re(\bar{r}')\bar{k}^- - \bar{k}^+).\bar{n}'\right]\exp\left(i\bar{k}^-.\bar{r}'\right) \tag{5.28}$$

where

$$\bar{k}^- = \bar{k}_0 - \bar{k}_s; \quad \bar{k}^+ = \bar{k}_0 + \bar{k}_s$$

The incident and scattered wave vectors are given from Figure 5.7 by

$$\bar{k}_0 = k\left(\hat{i}\sin\theta_i - \hat{k}\cos\theta_i\right)$$
$$\bar{k}_s = k\left(\hat{i}\sin\theta_s\cos\phi_s + \hat{j}\sin\theta_s\sin\phi_s + \hat{k}\cos\theta_s\right) \tag{5.29}$$

In the simplest cases, $\Re = 1$ (Neumann boundary condition) and $\Re = -1$ (Dirichlet boundary condition), one or the other of the terms within the integral in Equation 5.22 vanishes. For the case of scalar (acoustic) waves, these correspond to scattering by hard and soft boundaries, respectively. In the case of electromagnetic waves, the case $\Re = -1$ corresponds to the scattering of an s-polarized wave from a perfectly conducting surface.

The integral in Equation 5.28 may be approximated over the mean plane of the surface, provided that the surface slopes are not too steep

$$\bar{n}'dS \approx \left(-\hat{i}\frac{\partial h}{\partial X'} - \hat{j}\frac{\partial h}{\partial Y'} + \hat{k}\right)d^2R' \tag{5.30}$$

Here, $h(\bar{R}')$ is the local surface height measured from the mean plane. The scattered field may thus be expressed in the form

$$E(\bar{r}) = \frac{ik\exp(ikr)}{4\pi r} \int d^2R' \,\Theta(R')\left(a\,\partial h/\partial X' + b\,\partial h/\partial Y' - c\right) \times$$
$$\times \exp\left\{ik\left[AX' + BY' + Ch(\bar{R}')\right]\right\} \tag{5.31}$$

In this formula, $\Theta(\bar{R}')$ is an aperture function that is unity within the surface area, and $a\,b\,c\,A\,B\,C$ depend on the incident and scattering angles through the relations

$$
\begin{aligned}
&A = \sin\theta_i - \sin\theta_s\cos\phi_s & &a = \sin\theta_i(1-\mathfrak{R}') + \sin\theta_s\cos\phi_s(1+\mathfrak{R}') \\
&B = -\sin\theta_s\sin\phi_s & &b = \sin\theta_s\sin\phi_s(1+\mathfrak{R}') \\
&C = -(\cos\theta_i + \cos\theta_s) & &c = \cos\theta_s(1+\mathfrak{R}') - \cos\theta_i(1-\mathfrak{R}')
\end{aligned}
\tag{5.32}
$$

In order to progress further analytically, an additional assumption commonly made at this point is that the reflectivity is constant across the surface. This is a reasonable approximation if the surface slopes are small or if the surface constitutes the interface between two media of very different acoustic or dielectric properties so that $\mathfrak{R} \approx \pm 1$. The gradient terms in Equation 5.31 may then be eliminated through an integration by parts that results in the formula

$$E(\bar{r}) = -\frac{ik\exp(ikr)}{4\pi r} F(\theta_i,\theta_s,\phi_s) \int d^2R'\,\Theta(\bar{R}')\exp\left\{ik\left[aX' + bY' + ch(\bar{R}')\right]\right\} + e \tag{5.33}$$

where

$$F(\theta_i,\theta_s,\phi_s) = \frac{Aa}{C} + \frac{Bb}{C} + c \tag{5.34}$$

The final term, e, in Equation 5.33 is a contribution to the integral that arises at the edges of the surface area. This term is usually neglected, although there is some disagreement in the literature as to when this is a valid approximation [10].

In the case of normal incidence, $F = 2\mathfrak{R}$ and the exponent in the integral in Equation 5.33 may be expressed in the form

$$-\left\{i\bar{k}.\bar{R}'\sin\theta_s + ik(1+\cos\theta_s)h(\bar{R}')\right\} \tag{5.35}$$

In this case then, apart from an obliquity factor, Equation 5.33 is exactly of the same form as Equation 5.16 for scattering by a random phase screen into the Fraunhofer region with $\phi = -(1 + \cos\theta_s)kh$ and $A = \Theta$. The more general case of nonnormal incidence is also matched by a generalization of Equation 5.16. Thus, it has been demonstrated that the tangent plane approximation, together with a few additional simplifying assumptions, reduces the mathematics of scattering by a rough surface to that of scattering by a random phase screen.

It should be emphasized that the above theory, though widely used, is subject to a number of limitations that are analyzed in some detail in Reference [10]. In qualitative terms, it is a physical optics or short wave approximation requiring the local surface radius of curvature to be much greater than the wavelength. It takes no account of shadowing that is important at high angles of incidence. It takes no account of the rescattering of rays by other parts of the surface after their first encounter. Nevertheless, the results of numerical simulation [12] indicate that its range of validity is rather wider than these limitations might suggest.

References

1. Booker, H.G., J.A. Ratcliffe, and D.H. Shinn. Diffraction from an Irregular Screen with Applications to Ionospheric Problems. *Phil. Trans. Roy. Soc.* 242 (1950): 579–609.

2. Zardecki, A. Statistical Features of Phase Screens from Scattering Data. In *Topics in Current Physics 9*, in *Inverse Source Problems in Optics*. Ed. H.P. Baltes Berlin: Springer-Verlag, 1978.

3. Mercier, R.P. Diffraction by a Screen Causing Large Random Phase Fluctuations. *Proc. Camb. Phil. Soc.* A58 (1962): 382–400.

4. Salpeter, E.E. Interplanetary Scintillation I: Theory. *Astrophys. J.* 147 (1967): 433–48.

5. Jakeman, E., and P.N. Pusey. Non-Gaussian Fluctuations in Electromagnetic Radiation Scattered by a Random Phase Screen I: Theory. *J. Phys. A: Math. Gen.* 8 (1975): 369–91.

6. Taylor, G.I. Statistical Theory of Turbulence. *Proc. Roy. Soc.* A164 (1938): 476–90.

7. Pusey, P.N., and E. Jakeman. Non-Gaussian Fluctuations in Electromagnetic Radiation Scattered by a Random Phase Screen II: Application to Dynamic Scattering in a Liquid Crystal. *J. Phys. A: Math.Gen.* 8 (1975): 392–410.

8. Jakeman, E., and J.G. McWhirter. Correlation Function Dependence of the Scintillation behind a Deep Random Phase Screen. *J. Phys. A: Math. Gen.* 10 (1977): 1599–1643.

9. Jakeman, E., and W.T. Welford. Speckle Statistics in Imaging Systems. *Opt. Commun.* 21 (1977): 72–79.

10. Ogilvy, J.A. *Theory of Wave Scattering from Random Rough Surfaces*. Bristol: Adam Hilger, 1991.

11. Beckmann, P., and A. Spizzichino. *The Scattering of Electromagnetic Waves from Rough Surfaces.* Oxford: Pergamon, 1963. Reprinted Norwood, MA: Artech House, 1987.

12. see, for example, Thorsos, E.I., and D.R. Jackson. Studies of Scattering Theory Using Numerical Methods. in *Modern Analysis of Scattering Phenomena.* Eds. D. Maystre and J.C. Dainty. Bristol: IoP Publishing, 1991.

6

Scattering by Smoothly Varying Phase Screens

6.1 Introduction

It was shown in the last chapter that, under certain conditions, radiation scattered by a random phase-changing screen has the same Gaussian statistical properties as those predicted by the particle model of Chapter 4 and that under these special conditions the scattered field is effectively a random walk of many independent contributions. However, the random phase screen model also makes quantitative predictions for the statistical properties of the scattered wave field in many non-Gaussian Fresnel and Fraunhofer configurations that are commonly encountered in practical applications. A study of these non-Gaussian regimes provides many useful insights and has proved to be a valuable aid to the development of non-Gaussian models for the statistics of scattered waves.

In order to progress such a study it is necessary to make some assumptions about the statistical properties of the phase fluctuations introduced by the screen. In real applications, even if the phase screen model is a reasonable one for the scenario of interest, these properties are usually not well characterized, or are only known for a range of conditions limited by various practical considerations. This is often the case when measurements take place in the natural environment. For example, phenomena such as atmospheric turbulence or sea surface roughness vary widely according to the prevailing meteorological conditions and the proximity of fixed objects such as terrain, coastlines, and so forth. In view of these uncertainties, the methodology adopted has been to identify robust qualitative features of the scattered wave statistics that are associated with particular classes of phase screen models. In the present book, attention will be largely confined to deep screens (i.e., those that introduce path differences greater than the radiation wavelength) since it is these that generate the largest non-Gaussian intensity fluctuations and present the most severe limitations on system performance.

According to Equation 5.6 and Equation 5.8 of the last chapter, in order to evaluate the moments and correlation properties of radiation that have been scattered by a random phase screen, it is necessary to adopt a model for the joint characteristic function of the phase fluctuations. The choice for this role has in the past been almost exclusively the joint Gaussian process, although a few alternative models have been investigated in more recent years (Reference [1] contains a useful list of papers on this topic, but see also Chapter 8). The Gaussian process has the merit that its joint characteristic function has a simple analytic form requiring only specification of the first-order phase correlation function or spectrum (see, for example, Equation 2.10 and Equation 2.11). However, it is not necessarily an accurate model for all situations encountered in practice, and in Chapter 8 it will be shown that other models can lead to qualitatively different behavior of the scattered wave. Nevertheless, for the present, the joint Gaussian statistical model for the phase fluctuations will be adopted and an investigation will be carried out of the relationship between the properties of the scattered radiation and the form of the phase spectrum. This chapter will concentrate on *smoothly varying* screens characterized by a phase variable that is (spatially) differentiable to all orders. Note that by adopting the Taylor "frozen flow" hypothesis (Section 5.3), only consideration of properties implied by the spatial structure of the screens will be necessary in the Fresnel region, and the time variable will generally be omitted from the notation.

6.2 "Smooth" Models for the Phase Correlation Function

The earliest Gaussian phase screen model to be studied theoretically in a rigorous manner was one that also had a Gaussian (spatial) spectrum [2–11]. The spatial phase correlation function for the circularly symmetric case may then be written in the form

$$\langle \varphi(0)\varphi(\bar{r}) \rangle = \varphi_0^2 \rho(r) = \varphi_0^2 \exp\left(-r^2/\xi^2\right) \tag{6.1}$$

Two important features of model Equation 6.1 should be noted: (1) all its even power derivatives with respect to r near $r = 0$ are finite, and (2) it is characterized by a single length scale. By virtue of the theorem of Equation 2.34, the first property implies that the phase variations introduced by the screen are continuous and differentiable to all orders. The phase is then described as *smoothly varying*. The existence of only a single scale parameter in the model ensures that the phase fluctuations become uncorrelated (and statistically independent, since they are Gaussian) over distances of order ξ. It is of course possible to construct other spectral models with these features, but it can be shown that the properties of the scattered radiation are *qualitatively* similar for any deep Gaussian phase screen with a correlation function that can be expanded in the form

$$\rho(r) = 1 + \sum_{n=1}^{\infty} a_n \left(r/\xi\right)^{2n} \tag{6.2}$$

provided that the coefficients a_n are not too dissimilar. Thus the Lorentzian correlation function $\rho(r) = (1 + r^2/\xi^2)^{-1}$ also belongs to the smoothly varying, single scale class of models while the coefficients a_1, a_2 in the expansion (Equation 6.2) for the correlation function

$$\rho(r) = (1-b)\exp(-r^2/\xi^2) + b\exp(-r^2/a^2\xi^2) \tag{6.3}$$

may have widely different values, and this model can be used to characterize a smoothly varying medium with two distinct scales [11].

6.3 Qualitative Features of Scattering by Smoothly Varying Phase Screens

The most important factor governing the properties of fluctuations in the scattered radiation is in fact the "smoothness" of the scattered wave front. For smoothly varying wave fronts (whether they are Gaussian or not), constructing rays can delineate the energy flow as illustrated in Figure 6.1. This geometrical optics approximation reveals some of the qualitative features of this kind of scattering. It can be seen that as the rays leave the screen they begin to converge or diverge as a result of the refractive effect of the phase inhomogeneities. This leads to increasing variations in intensity. Eventually random focusing events occur. These are associated with individual lens-like features of the screen and it is evident that the intensity fluctuations in

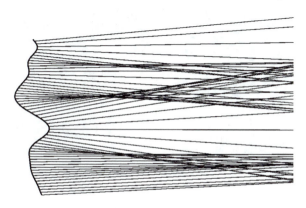

FIGURE 6.1
Ray propagation beyond a smoothly varying phase screen.

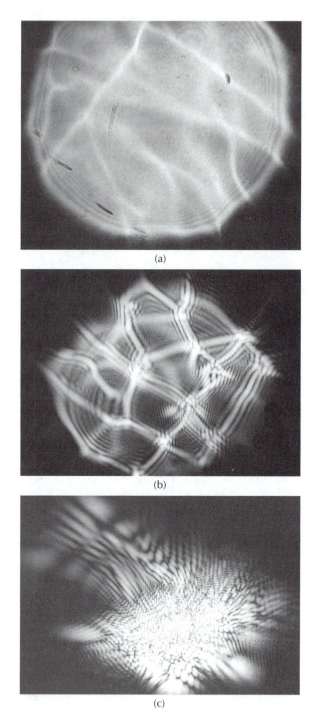

(a)

(b)

(c)

FIGURE 6.2
Intensity patterns generated when laser light is passed through a turbulent plume of air (a) close to the plume, (b) in the focusing region, (c) far beyond the plume.

their neighborhood will be particularly large. From the focusing points, caustics, that is, regions where rays overlap, propagate into the far field; these are mathematical singularities or *catastrophes* in the scattered wave field where the ray density becomes "infinite" [12]. However, in practice they are broadened by diffraction as a result of the finite radiation wavelength. This leads to the appearance of fringes that are clearly visible in the optical patterns shown in Figure 6.2. In the case of a plane incident wave, beyond the focusing region the caustics themselves begin to overlap, and here interference between the contributions from independent elements of the screen begins to occur. Eventually, so many elements contribute that the statistics of the scattered wave approach the Gaussian limit as discussed in the last chapter.

Thus when a plane wave propagates into the Fresnel region through a smoothly varying deep random phase screen, we expect intensity fluctuations to build up to a maximum in the region of focusing and then decrease again, reaching the Gaussian speckle limit in the far field. To be specific, the scintillation index will increase from zero at the screen to a maximum value in excess of unity in the focusing region and then decrease to unity at long propagation distances. This behavior was first predicted theoretically by Mercier in 1962 [2] and first measured in a series of laboratory optical frequency experiments reported by Parry et al. in 1977 [13].

6.4 Calculation of the Scintillation Index in the Fresnel Region

Calculation of the scintillation index from Equation 5.8 is surprisingly difficult. After making a Gaussian assumption for the statistics of the phase fluctuations, the Fresnel region Equation 5.8 reduces when $\Delta = 0$ to

$$S^2 + 1 = \frac{1}{\lambda^2 z^2} \int d^2 r \, d^2 r' \exp\left(-\varphi_0^2 F(\vec{r}, \vec{r}') - \frac{ik}{z} \vec{r}.\vec{r}' \right) \tag{6.4}$$

where

$$F(\vec{r}, \vec{r}') = 2 - 2\rho(\vec{r}) - 2\rho(\vec{r}') + \rho(\vec{r} + \vec{r}') + \rho(\vec{r} - \vec{r}') \tag{6.5}$$

Here ρ is the correlation function of the phase fluctuations, and for a smoothly varying model will be given by an expansion of the form of Equation 6.2. It is not difficult to show that when $\varphi_0^2 \gg 1$ the integrand is exponentially small except where the magnitudes r or r' are small. The analytical properties of the integrand can be illustrated most simply for a

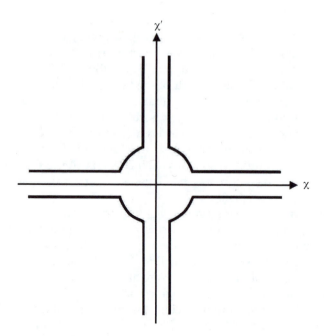

FIGURE 6.3
Principal region of integration contributing to Equation 6.4 in the case of a corrugated screen.

screen that is corrugated in the x-direction only, where Equation 6.4 reduces
to an integral over the x - x' plane. Figure 6.3 shows a diagram of this plane
highlighting the regions from which the main contributions arise. Near to
the screen the contribution from the central area, where x and x' are *both*
small, dominates the integral. This approaches the value of unity as $z \rightarrow 0$
so that $S^2 = 0$. Far from the screen, as $z \rightarrow \infty$, contributions from regions
along the axes dominate and they each give a value of unity so that the
scintillation index is also unity, as would be expected if the scattered field
was a Gaussian process. The behavior between these limiting situations
cannot be evaluated exactly by analytical methods. Indeed, until computing
power expanded in the early nineteen eighties, it was also difficult to calculate numerically in the deep phase screen case because of the oscillatory term
in the integrand of Equation 6.4 and the singular behavior of the exponential
factor. However, an approximate analytical result was obtained by a combination of steepest descents and function modelling techniques that will now
be described [11].

In the case of an isotropic screen where the scintillation index is given by
Equation 6.4, the principal contributions to the integral arise from regions
where r or r' and hence $F(r, r')$ is small. The integral possesses twofold
symmetry in the r - r' plane and may be expressed in terms of contributions
from the region bounded by the r-axis and the line $r = r'$. Since $\phi_0^2 \gg 1$, a
steepest descents approximation can be used to reduce the integral over r
in this region by expanding $F(r, r')$ in powers of r using the series

$$\rho(\vec{r}) = 1 + \tfrac{1}{2}(\vec{r}.\vec{\nabla})^2 \rho(0) + \tfrac{1}{24}(\vec{r}.\vec{\nabla})^4 \rho(0) + O(r^6) \tag{6.6}$$

This leads to an expression of the form

$$S^2 + 1 = \frac{4\pi}{\lambda^2 z^2} \int\limits_0^\infty r\,dr \int\limits_0^r r'dr' \int\limits_0^{2\pi} d\psi \cos(krr'\cos\psi) \times$$

$$\times \exp\left\{-\phi_0^2 r'^2 \left[D^2\rho(r) - D^2\rho(0) + r^2\cos^2\psi D^4\rho(r)\right]\right\} \tag{6.7}$$

Here $D^2 \equiv (1/r)d/dr$. This approximate formula which is extremely accurate, even for moderate values of ϕ_0^2, greatly facilitates numerical computation when ϕ_0^2 is large. Figure 6.4 shows the results of a calculation in which the single scale model (Equation 6.1) is used in Equation 6.7.

It is possible to make further progress analytically by performing some fairly crude but simple modelling of the exponent in Equation 6.7. Although the formula obtained in this way is obviously less accurate than the computed results, it is relatively simple and provides some useful insight into

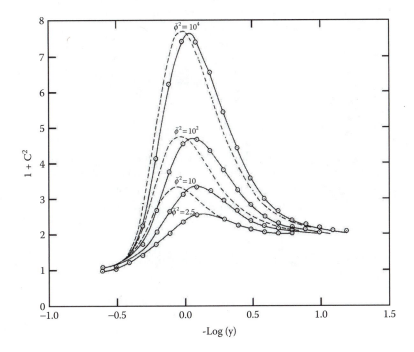

FIGURE 6.4
Scintillation index in the Fresnel region plotted as a function of the logarithm of the normalized distance y from the screen, o-o-o-o computed results, - - - - - - analytical Equation 6.11 [11].

the structure of the scintillation curves. In the central region $r \le \alpha$, where $\alpha = [-2D^2\rho(0)/D^4\rho(0)]^{1/2}$, the approximation

$$D^2\rho(r) \approx D^2\rho(0) + \left(r^2/2\right)D^4\rho(0) \tag{6.8}$$

is used with $r^2 D^4\rho(r) \approx r^2 D^4\rho(0)$. Assuming that $t = [\phi_0 D^2\rho(0)]^2/D^4\rho(0) \gg 1$ the contribution from this region is given to a very good approximation by

$$I_1 = -\frac{\pi}{\lambda^2 z^2} \int_0^\pi d\psi \int_0^\infty dx \ln\left(\frac{x}{\alpha^4}\right) \cos\left(kx^{1/2}z^{-1}\cos\psi\right) \times$$
$$\times \exp\left[-\tfrac{1}{2}\phi_0^2 t D^4\rho(0)(1 + 2\cos^2\psi)\right] \tag{6.9}$$

This is the non-Gaussian or single scatterer contribution to the scintillation index. It decreases asymptotically to zero for large values of z when many correlation areas of the screen contribute to the scattered intensity pattern. For $r > \alpha$ it will be assumed that $D^2\rho(r) \approx D^4\rho(r) \approx 0$ and provided $t \gg 1$ the upper limit of the r' integral in Equation 6.7 may be extended to infinity without incurring serious error. The contribution from this second region then reduces to the form

$$I_2 = \frac{-k^2}{\phi_0^2 z^2 D^2\rho(0)} \int_\alpha^\infty rdr \exp\left(k^2 r^2/4z^2\phi_0^2 D^2\rho(0)\right) \tag{6.10}$$

This term represents the effect of interference and is associated with the presence of speckle in the intensity pattern. It saturates at the Gaussian value of two for large propagation distances, but is negligible close to the screen where the number of scattering centers (correlation cells) is on average much less than one. The value of I_2 is easily evaluated and by carrying out a partial integration of Equation 6.9 it is possible to express the scintillation index in the form

$$S^2 + 1 = 2\exp(-3y^2) + \sqrt{3}y^2 \exp(-y^2)\left[\ln\left(2\gamma t/3\right) + R(y)\right] \tag{6.11}$$

where

$$y^2 = k^2/\left(6z^2\phi_0^2 D^4\rho(0)\right)$$

$$R(y) = \int_0^1 \frac{dx}{x}\left\{\frac{\exp(y^2 x)}{\sqrt{1+2x}} - 1\right\} \tag{6.12}$$

Here $\gamma = \exp(c)$, c is the Euler-Masheroni constant. For the Gaussian model (Equation 6.1), $D^2\rho(0) = -2/\xi^2$, $D^4\rho(0) = 4/\xi^4$, and $t = \varphi_0^2$.

6.5   Predicted Behavior of the Scintillation Index in the Fresnel Region

Figure 6.4 compares a plot of the approximate Equation 6.11 as a function of log $(1/y)$ with the results of a more accurate numerical calculation [11]. It can be seen that the accuracy of the analytical approximation improves as the phase variance increases. This is only to be expected since the steepest descents method then becomes a better approximation. Equation 6.11 predicts an increase of the scintillation index from zero at the screen when $y \gg 1$ to a maximum value near $y = 1$, followed by a slow decline to unity when $y \ll 1$ corresponding to large propagation distances. A noteworthy feature of this deep phase screen result is that, whatever smooth, single-scale, phase correlation function model is used, it depends on only two parameters: the effective variance t of the phase fluctuations introduced by the screen, and the dimensionless propagation distance $1/y$. The latter parameter assumes particular importance when the phase fluctuations are expressed in terms of path differences introduced into the incident wave, that is, $\varphi_0 = kh_0$, where h_0 is a root mean square (rms) path variation. It is then found that

$$y^{-1} = 2\sqrt{6}\,\frac{zh_0}{\xi^2} \tag{6.13}$$

Thus the position of the focusing peak near $y = 1$ is independent of wavelength and is indeed determined by the *geometry* of the initial wave front, being consistent with the notion of a wave front distortion of "height" h_0 and lateral extent ξ focusing near $z = \xi^2/h_0$, that is, near $y = 1$. Moreover, it can be argued that at any propagation distance z the area contributing to the intensity according to geometrical optics is of order $z^2 h_0^2/\xi^2$.

It is instructive to expand Equation 6.11 for the scintillation index in the asymptotic regime beyond the peak in powers of the parameter y. When the wave number is sufficiently large, the integral R may be neglected by comparison with the logarithmic term in this region, and to second order in y we obtain

$$S^2 + 1 \approx 2(1 - 3y^2) + \sqrt{3}y^2 \ln(2\gamma t/3) \tag{6.14}$$

The structure of Equation 6.14 is the same as the results of Equation 4.12, Equation 4.22, and Equation 4.24 for the random walk model if $N = 1/3y^2 = 8z^2 h_0^2 / \xi^4$. According to the simple geometrical argument given above, this is indeed the number of correlation cells of the screen contained in the area contributing to the intensity at propagation distance z. The first factor in Equation 6.14 that comes from the axial regions illustrated in Figure 6.3 is the contribution from interference. The final logarithmic term comes from the central region and is associated with the average contribution of a single phase inhomogeneity or correlation cell.

It is evident that the predictions of the single-scale model (Equation 6.1) are in qualitative agreement with the experimental data shown in Figure 6.5. However, the measured width of the peak in the scintillation plot is greater than the theoretical prediction. Since the data were generated by passing laser light through a thermal plume of air generated above a heating element [13], this is probably due to the presence of multiple scales — a characteristic often attributed to turbulent flows [14]. The result of calculations using the two-scale model (Equation 6.3) (shown as the solid line in Figure 6.5) gives better agreement with the data at short propagation distances and appears to support this conjecture, although the need for more fitting parameters detracts somewhat from the significance of the result. Moreover, fluid flow of this kind is usually characterized by a *cascade* of scale sizes and a *power-law* spectrum — a model with qualitatively different scattering characteristics that will be discussed in the next chapter.

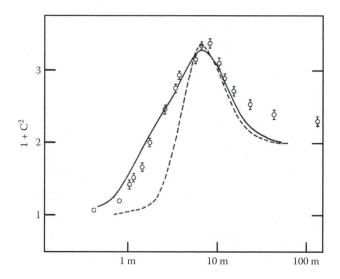

FIGURE 6.5

Scintillation index in the Fresnel region as a function of distance z from the screen, Φ experimental data, - - - - - - - single-scale model Equation 6.1, _____ two-scale model Equation 6.3 with a = 0.15, b = 0.01, ϕ_0^2 = 120 [11].

6.6   Higher-Order Statistics in the Fresnel Region

It is readily established that analytical calculation of the intensity fluctuation distribution or its moments higher than the second using the above direct approach is impractical in general. Salpeter [3] suggested that in a region governed by geometrical optics the distribution of intensity fluctuations has an inverse cube dependence on the intensity leading to divergence of the second intensity moment. His argument was based on a consideration of ray propagation near the screen and he observed that the divergence was removed by diffraction, resulting in the logarithmic term visible in Equation 6.11 and Equation 6.14. Later, some further progress was made in the high frequency limit using more sophisticated mathematical tools. Writing $t = k^2 h_0^2$ as above, it can be seen from the result of Equation 6.11 that the second intensity moment diverges like $\ln(k)$. It may be shown that intensity falls off inversely with the square root of distance away from a caustic [3] so that in the geometrical optics limit, where these features dominate the intensity pattern, it might be anticipated that all the higher moments of the intensity will diverge. Berry [15] showed how this divergence depends on k and related his results to the singularity classification of catastrophe theory. In particular for $n > 2$

$$\lim_{k \to \infty} \left\langle I^n \right\rangle \approx k^{\nu_n} \tag{6.15}$$

where ν_n is the nth critical exponent. The result in Equation 6.15 is valid for a phase screen that can be expressed as the sum of N random sinusoids. Berry proved that when $N \to \infty$, corresponding to the Gaussian screen discussed above, then the values of the critical exponents depend only on whether the wave front is distorted in one direction (i.e., corrugated) generating two-dimensional caustics, or whether these geometrical features are fully three-dimensional objects. Experimental measurements in which light is scattered from a rippled water surface have shown reasonable agreement with theoretical predictions [16]. However, the extreme short wave limit in which these results apply is difficult to achieve and measure experimentally and may rarely occur in practice.

Greater insight into the behavior of the wave field in the Fresnel region beyond a smoothly varying Gaussian phase screen is gained from computer simulation, described in Chapter 14. Figure 6.6 shows complex realizations of the scattered field produced by a corrugated phase screen. These are the variations along a line parallel to the x-axis (in practice these would be generated at a point detector by uniform transverse motion of the screen). As expected, near the screen the trajectory of the field vector is contained within a narrow annulus bounding the weak amplitude

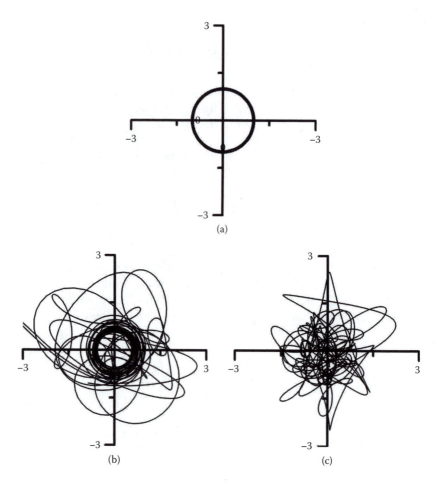

FIGURE 6.6
Numerical simulation of the complex wave field in the Fresnel region beyond a smoothly varying corrugated Gaussian phase screen (a) close to the screen, (b) in the focusing region, (c) in the region of interference. Here $\phi_0^2 = 400$ and $z/\xi =$ (a) 30, (b) 2000, and (c) 10,000.

fluctuations that occur in this regime. As the focusing region is reached, the annulus is still present, although broadened. However, the presence of the focusing peaks is apparent in the occasional large amplitude fluctuations (one of these goes outside the plotted region in Figure 6.6b); these are also associated with an increasing chance of the trajectory passing close to the origin, which is related to the formation of nulls in the intensity near the focusing peaks. Eventually, the field components are concentrated around the origin (Figure 6.6c), the annular structure disappears, and the characteristic Gaussian probability density begins to emerge. Plots of the intensity probability densities corresponding to these realizations (Figure 6.7) show the transition between the delta function behavior near the screen and the negative exponential behavior far beyond it, with the

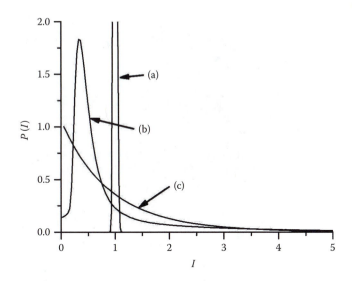

FIGURE 6.7
Intensity probability densities corresponding to the simulations shown in Figure 6.6.

focusing regime result showing a characteristic peak and a long tail continuing out to high values of intensity. In the region beyond the focusing peak, the class of K-distributions encountered in Chapters 3 and 4 have been found to provide a good fit to experimental data [17].

6.7   Spatial Coherence Properties in the Fresnel Region

A few results for the spatial intensity autocorrelation function of the radiation have also been derived for the region beyond the scintillation peak in Figure 6.4. Interest in the spatial coherence properties of radiation that has been scattered by a random phase screen dates back to the early days of ionospheric research [18,19]. Such properties have a diagnostic role as a means of remote sensing, with applications ranging from the measurement of stellar diameters to fluid flow [20,21]. Analytical calculation of the intensity correlation function in the deep phase screen limit is difficult. However, using the mathematical approximation and modelling methods detailed above, Jakeman and McWhirter [10] were able to show for the corrugated deep Gaussian random phase screen with Gaussian spectrum that the second-order transverse coherence function is given approximately by

$$g^{(2)}(\chi) = 1 + \left|g^{(1)}(\chi)\right|^2 - \frac{2q}{\sqrt{\pi}}\left[\frac{\sin\left(k\xi\chi/z\sqrt{3}\right)}{k\xi\chi/z\sqrt{3}} + \exp\left(-\frac{\xi^2\chi^2}{8z^2h_0^2}\right)\right] +$$

$$+ \frac{2q}{\sqrt{\pi}}\left[Ci\left(\frac{k\xi\chi}{z\sqrt{3}}\right) + \frac{1}{2}E_1\left(\frac{\xi^2\chi^2}{8z^2h_0^2}\right)\right] \tag{6.16}$$

where E_1 is the exponential integral and Ci the cosine integral [Ref 4, Ch 3]. Here, the spatial separation is $\bar{\Delta} \equiv (\chi, 0)$, $q = \xi^2/2zh_0\sqrt{6}$ and the field correlation function, according to Equation 5.6 is for $\varphi_0 \gg 1$

$$\left|g^{(1)}(\chi)\right| = \exp\left(-k^2h_0^2\chi^2/\xi^2\right) \tag{6.17}$$

The result (Equation 6.16) is only valid for small values of q. It reduces to the corrugated screen equivalent of Equation 6.14 when $\chi = 0$ and asymptotically to the Siegert Gaussian factorization theorem as $q \to 0$. Its general structure is similar to the result in Equation 4.29 of the random walk model, bearing in mind that in the corrugated phase screen case, $N \sim 1/q$ (see discussion following Equation 6.14). Note that when the screen is moving uniformly across the incident wave with transverse velocity v, the temporal coherence function may be obtained from Equation 6.16 by means of the transformation $\chi \to v\tau$. Three distinct length scales are evident in the formula. The smallest is the interference or speckle size ξ/φ_0 and is independent of propagation distance. The diffraction scale $z/k\xi$ and the geometrical optics scale zh_0/ξ both increase linearly with distance. Thus Equation 6.16 serves to confirm the heuristic description in terms of geometrical optics, diffraction, and interference effects given earlier in the chapter. Experimental measurement of scattering by a thermal plume in the asymptotic region where Equation 6.16 is valid also demonstrate conclusively that the scattered intensity pattern in real situations is characterized by more than one length scale (Figure 6.8) [17].

6.8   Calculation of the Scintillation Index in the Fraunhofer Region

When the area of the phase screen contributing to the scattered intensity at a point is limited by the size of the illuminated area rather than by Fresnel zone cancellation effects or by tilts in the scattered wave front, the scattered field can be expressed in the Fraunhofer form (Equation 5.16). In order to make analytical progress it will be assumed here that the screen is placed at

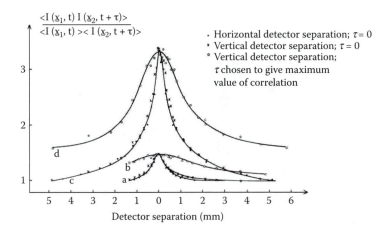

FIGURE 6.8

Laser light scattering by a thermal plume, measured cross correlation between two separated detector outputs in the focusing region, delay time τ [17].

the waist of a Gaussian beam of width W (Equation 4.3). After making the further assumption of Gaussian phase statistics, it is possible to reduce the expression for the scintillation index in this case to [11]

$$S^2 + 1 =$$

$$= \frac{16\pi W^2 \int d^2r' d^2r'' d^2r''' \exp\left[-\varphi_0^2 G(\bar{r}', \bar{r}'', \bar{r}''') + 2i\bar{k}.r'' \sin\theta - (r'^2 + r''^2 + r'''^2)/W^2\right]}{\left[\int d^2r' d^2r'' \exp\left\{-\varphi_0^2\left[1 - \rho(\bar{r}')\right] + i\bar{k}.\bar{r}' \sin\theta - (r'^2 + r''^2)/2W^2\right\}\right]^2}$$

(6.18)

where

$$G(\bar{r}', \bar{r}'', \bar{r}''') = 2 - \rho(\bar{r}'' + \bar{r}''') - \rho(\bar{r}'' - \bar{r}''') - \rho(\bar{r}'' + \bar{r}') - \rho(\bar{r}'' - \bar{r}') +$$

$$+ \rho(\bar{r}' + \bar{r}''') + \rho(\bar{r}' - \bar{r}''')$$

It was argued in the last chapter that the integral in Equation 6.18 reduces to unity when the illuminated area is so small that the scattered field consists of a single large diffraction lobe, and to two when the area contains many independent scattering elements. In between these two limiting situations, the integrals are difficult to evaluate. However, as in the Fresnel region case, progress can be made using the method of steepest descents, together with some function modelling. Again, it is easier to comprehend the nature of the integrand in the corrugated screen case when Equation 6.18 reduces to a triple integral. Figure 6.9 highlights the regions in the integration volume

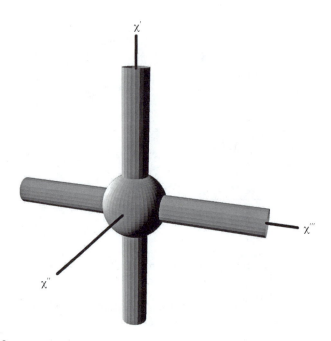

FIGURE 6.9
Principal region of integration contributing to Equation 6.18 in the case of a corrugated screen.

where the major contributions to the integral reside. These are confined near the $x'' = 0$ plane where either x' or x''' are small.

The denominator of Equation 6.18 can be evaluated using a steepest descents approximation when $\phi_0 \gg 1$ with the result

$$\left(4\pi^2 W^2/f_0\right)^2 \exp\left(-k^2 \sin^2\theta/f_0\right) \tag{6.19}$$

In this expression $f_0 = W^{-2} - \phi_0^2 D^2 \rho(0)$. The principal contribution to the numerator derives from a region where r'' is small and the following Taylor expansion is used

$$G(\vec{r}', \vec{r}'', \vec{r}''') \approx F(\vec{r}', \vec{r}''') - \left[r''^2 D^2 + \left(\vec{r}''.\vec{r}'''\right)^2 D^4\right]\rho(r''') - $$
$$- \left[r''^2 D^2 + \left(\vec{r}''.\vec{r}'\right)^2 D^4\right]\rho(r') \tag{6.20}$$

where F is defined by Equation 6.5. The r'' integral can then be performed exactly and the resulting expression evaluated by following the procedure described in Section 6.4. Thus a steepest descents approximation employing the approximation of Equation 6.6 leads to

$$16\pi^2 W^2 \int_0^{2\pi} d\psi \int_0^\infty r\,dr \int_0^r r'dr' f^{-1}(r,\psi) \exp\left[-k^2 \sin^2 \theta / f(r,\psi)\right] \times$$

$$\times \exp\left\{-\phi_0^2 r'^2 \left[D^2\rho(r) - D^2\rho(0) + r^2 \cos^2 \psi D^4\rho(r)\right] - \left(r^2 + r'^2\right)/W^2\right\}$$

(6.21)

Here

$$f(r,\psi) = W^{-2} - \phi_0^2 \left[D^2\rho(r) + D^2\rho(0) + r^2 \cos^2 \psi D^4\rho(r)\right]$$

This expression is evaluated by setting $\cos^2 \psi \approx \frac{1}{2}$ and dividing the r-integral into two regions accordingly as r is less than or greater than $\alpha/\sqrt{2}$, defined in Section 6.4. Using the same modelling procedure and taking $f(r,\psi) \approx f_0 - \phi_0^2 D^2\rho(0)$ for simplicity, the contribution to Equation 6.21 from the region $\lfloor 0, \alpha/\sqrt{2} \rfloor$ may be written approximately [16]

$$\frac{16\pi^4 W^2}{\phi_0^2 D^4\rho(0)} \frac{\exp\left\{\left[\phi_0^2 W^4 D^4\rho(0)\right]^{-1} - k^2 \sin^2 \theta / \left[f_0 - \phi_0^2 D^2\rho(0)\right]\right\}}{\left[f_0 - D^2\rho(0)\right]} \times$$

$$\times \left\{E_1\left(\left[\phi_0^2 W^4 D^4\rho(0)\right]^{-1}\right) - 2E_1\left(\left[\phi_0^2 W^4 D^4\rho(0)\right]^{-1} + \alpha^2/2W^2\right)\right\}$$

(6.22)

In the second region of integration $\lfloor \alpha/\sqrt{2}, \infty \rfloor$ set $D^2\rho(r) - D^2\rho(0) + r^2 \cos^2 \psi$

$D^4\rho(r) \approx -D^2\rho(0)$ so that $f(r,\psi) = f_0$. Assuming that $\exp\left[\phi_0^2 \alpha^2 D^2\rho(0)\right]$ is negligible for $\phi_0^2 \gg 1$, the following contribution is obtained from this region

$$\left(32\pi^4 W^4/f_0^2\right) \exp\left(-\alpha^2/2W^2 - k^2 \sin^2 \theta/f_0\right)$$

(6.23)

Combining the results of Equation 6.22 and Equation 6.23 leads finally to [11]

$$S^2 + 1 = 2\exp(-s) + \frac{s}{t}\frac{(s+t)^2}{(s+2t)}\left[E_1\left(\frac{s^2}{t}\right) - 2E_1\left(\frac{s^2}{t}+s\right)\right]\exp\left[\frac{s^2}{t} + \frac{k^2 W^2 \sin^2 \theta}{(s+t)(s+2t)}\right]$$

(6.24)

where $s = -D^2\rho(0)/W^2 D^4\rho(0) = \xi^2/2W^2$ and $t = \phi_0^2$ in the case of a Gaussian spectrum.

6.9 Predicted Behavior of the Scintillation Index in the Far Field

It is not difficult to check that Equation 6.24 approaches the expected values when W is very large or very small compared to ξ. A plot of the function with $\theta = 0$ for various values of t is shown in Figure 6.10 [11]. Note that apart from the angle dependence there are again only two significant parameters in the result whatever even-powered, single-scale phase correlation function is used. The first term in Equation 6.24 is associated with interference. It comes from the axial regions of Figure 6.9 and increases to two when s is small corresponding to a large aperture. The remaining terms come from the central region of Figure 6.9 and describe the contribution of the diffraction-broadened caustic associated with an individual phase correlation cell. These terms decrease to zero when s is small, but take the value of unity when it is large. In between these two limiting cases the scintillation index peaks in a fashion that is rather similar to the Fresnel result. This superficial resemblance is misleading, however, since the position and height of the peak is governed by diffraction as well as by geometrical optics effects. For large ϕ_0 the peak is of magnitude $\propto \phi_0$ (cf., Fresnel case $\ln\phi_0$) and occurs near $s^2 = t$,

that is, $W \approx \phi_0^{-1/2} \left[D^4\rho(0)\right]^{-1/4}$. This is when the speckle or aperture diffraction scale size becomes comparable with the geometrical optics scale associated with local curvature of the scattered wave front and is wavelength dependent, unlike the scintillation peak in the Fresnel region. The scintillation index of laser light scattered from a ruffled water surface in laboratory experiments have been found to exhibit this behavior (Figure 6.10) [22].

It is also instructive to expand Equation 6.14 in the asymptotic regime beyond the peak where it is found that

$$S^2 + 1 = 2(1-s) + \frac{s}{2}\ln(\gamma t)\exp\left(\frac{k^2 W^2 s \sin^2\theta}{2t}\right) \tag{6.25}$$

The result in Equation 6.25 is again of the random walk form (Equation 4.12) when s is identified with N^{-1}. For the Gaussian spectral model, this is an accurate analogy since $s^{-1} = 2W^2/\xi^2$ and is therefore of the order of the number of phase correlation areas in the illuminated area. The scintillation index of laser light scattered from electrohydrodynamic turbulence in layers

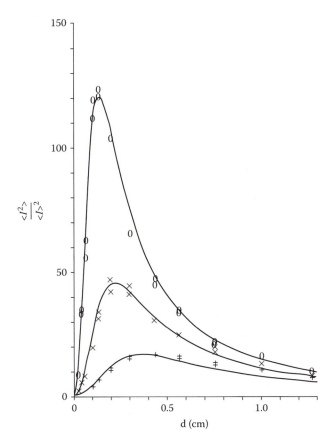

FIGURE 6.10
Scintillation in the far field, _____ normalized second intensity moment as a function of beam diameter d predicted by Equation 6.24 with $\xi = 2.6$ cm and $\phi_0 = 50, 135$, and 360 compared to experimental measurements of laser light scattered through a ruffled water surface [22].

of nematic liquid crystals shown in Figure 6.11 [23] exhibit the implied inverse dependence on illuminated area.

The final angle dependent factor in Equation 6.25 may, for the case of a Gaussian spectral model, be written as $\exp(\sin^2 \theta / 4m_0^2)$, where the rms slope of the scattered wave front is $m_0 = h_0/\xi$. This factor is inversely proportional to the probability of finding slopes on the wave front that face the observation direction, and therefore takes account of the reduced chance of finding caustics at large scattering angles. The same property also dictates the behavior of the mean intensity that appears squared in the denominator on the right-hand side of Equation 6.18

$$\langle I \rangle \propto \int d^2\bar{r} \exp\left\{-\varphi_0^2\left[1-\rho(\bar{r})\right]+i\bar{k}.\bar{r}\sin\theta - r^2/2W^2\right\} \approx \exp\left[-\sin^2\theta/4m_0^2\right]$$

$$(6.26)$$

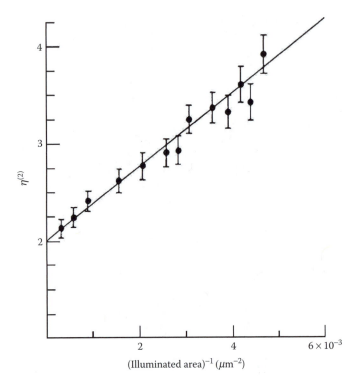

FIGURE 6.11

Scintillation in the far field. Experimental measurements of fluctuations in laser light scattered by electrohydrodynamic turbulence in a liquid crystal showing the second normalized intensity moment inversely proportional to illuminated area [23].

Thus the mean intensity is directly proportional to the probability of finding an element of the scattered wave front normal to the direction of view, while in the non-Gaussian regime the deviation of the scintillation index from unity is *inversely* proportional to this factor and increases rapidly with angle for smoothly varying Gaussian phase screens. This behavior was observed in the liquid crystal experiments [23] and similar trends have also been observed in light scattering through a turbulent mixing layer of brine and water [24] as well as in microwave scattering from the sea surface [25].

6.10 Higher-Order Statistical Properties in the Far Field

It is not possible to extend the analytical procedures outlined above to calculate higher intensity moments and correlation functions or the probability density function of intensity fluctuations itself. Recently, numerical simulations of the effect of passing a Gaussian random phasor with Gaussian spectrum through a Lorentzian filter have been carried out [26] using techniques described in Chapter 14. The resulting signal may be written

$$S(t) = \lambda \int_{-\infty}^{t} dt' \exp\left[i\phi(t') + \lambda(t' - t) + i\omega(t' - t) \right] \tag{6.27}$$

Comparison with Equation 5.16 suggests that this configuration is mathematically equivalent to scattering into the far field by a smoothly varying corrugated phase screen illuminated by a beam with negative exponential profile. It therefore gives some indication of the expected statistical properties in the scattering problem. Figure 6.12 shows realizations of the initial phase fluctuation and the intensity variations generated as the "beam width" λ^{-1} is increased. Remarkably, the individual diffraction broadened caustics are clearly visible at intermediate values. The associated distributions are shown in Figure 6.13. Note the regime (Figure 6.13a) where a double hump is predicted. There is little experimental evidence for this, probably because the scattering media studied contained inhomogeneities of more than one size. On the other hand, in the asymptotic regime, where the deviation of the scintillation index from unity is inversely proportional to the illuminated area, the distribution acquires a characteristic high tail lying above that of the negative exponential distribution. Under these conditions, experimental data is usually found to lie very close to the K-distribution, Figure 6.14 [27].

6.11 Coherence Properties in the Far Field

As in the Fresnel geometry, some formulae for the Fraunhofer region spatial coherence functions have been derived for a corrugated Gaussian phase screen with Gaussian spectrum in the asymptotic regime beyond the peak in the scintillation plot. For small angles θ symmetrically placed about the axis, it may be shown that the normalized first-order correlation function is given by

$$g^{(1)}(V) = \exp\left(-k^2 V^2 W^2 / 8\right) \tag{6.28}$$

while the normalized intensity correlation function takes the form [11]

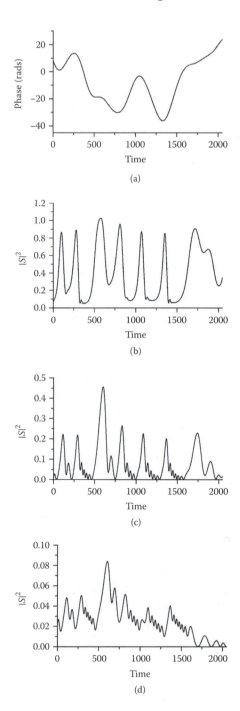

FIGURE 6.12
Lorentzian filtered, smoothly varying Gaussian random phasor Equation 6.27, numerically simulated time series (a) phase, (b–d) intensity with $\lambda = 0.05$, 0.01, and 0.002, respectively. The correlation time of the Gaussian phasor is 200 units.

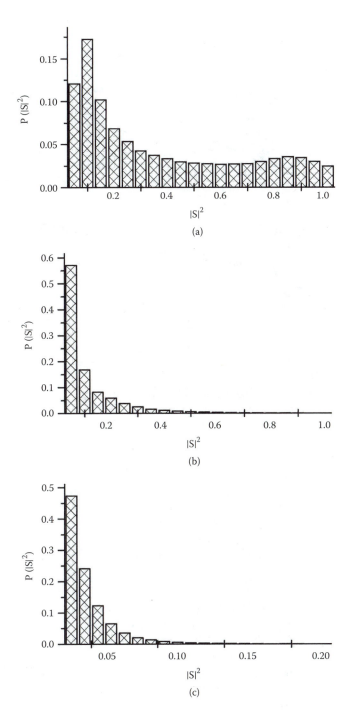

FIGURE 6.13
Probability distributions of intensity for the time series illustrated in Figure 6.12.

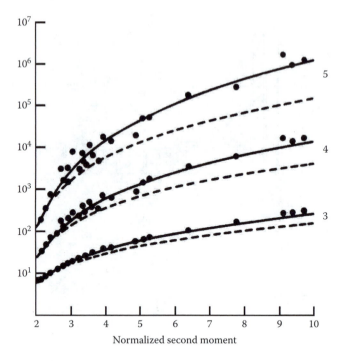

FIGURE 6.14
Higher normalized intensity moments of the K-distribution Equation 3.26 plotted as a function of the second, comparison with experimental measurements of fluctuations in laser light scattered by electrohydrodynamic turbulence in a liquid crystal [27].

$$g^{(2)}(V) = 1 + \left| g^{(1)}(V) \right|^2 - \frac{2\xi}{W\sqrt{3\pi}} \exp\left(\frac{V^2}{8m_0^2} \right) \left[\exp\left(\frac{V^2}{8m_0^2} \right) + \frac{\sin\left(kV\xi/\sqrt{3} \right)}{kV\xi/\sqrt{3}} \right] +$$

$$+ \frac{2\xi}{W\sqrt{6\pi}} \exp\left(\frac{V^2}{8m_0^2} \right) \left[Ci\left(\frac{kV\xi}{\sqrt{3}} \right) + \frac{1}{2} E_1\left(\frac{V^2}{8m_0^2} \right) \right]$$

$$(6.29)$$

Here $V = 2 \sin \theta$. When W is large compared to the phase inhomogeneities, so that many independent elements of the screen contribute to the scattered field, Equation 6.29 reduces to the Siegert relation (Equation 2.25). When $V = 0$, it reduces to the forward scattering ($\theta = 0$) form of Equation 6.25 applicable to a corrugated phase screen. Three angular widths are discernible in Equation 6.29. One of order $(kW)^{-1}$ is due to interference and is manifest in the first-order correlation function. Another is associated with diffraction from individual phase inhomogeneities of order $(k\xi)^{-1}$, and one reflects the geometrical properties of the scattered wave front, being proportional to the

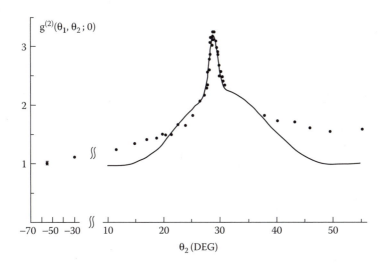

FIGURE 6.15
Intensity spatial coherence function of laser light scattered by electrohydrodynamic turbulence in a liquid crystal [17].

rms slope m_0. In the asymptotic region, $W \gg \xi$, and so the smallest features in the scattered intensity pattern are due to interference. On the other hand, the largest features arise from the geometry of the wave front since $kWm_0 = \varphi_0 W/\xi \gg 1$. Experimental evidence for the presence of at least two distinct length scales in Fraunhofer non-Gaussian intensity patterns has been obtained from measurements of laser light scattered from electrohydrodynamic turbulence in layers of nematic liquid crystals shown in Figure 6.15 [23].

It has already been noted in the last chapter that, in the Fraunhofer region, uniform motion of the phase screen leads to speckle evolution as well as translation if wave front curvature is present. This is most easily demonstrated through a calculation in the Gaussian regime where the Siegert factorization theorem requires evaluation of only the first-order correlation function. The most transparent results are obtained in a paraxial approximation where Equation 5.2 and Equation 5.5 are generalized to include a Gaussian beam profile. In the case of a Gaussian phase screen, the field correlation function is proportional to

$$\int d^2\bar{r}d^2\bar{r}' \exp\left[-\varphi_0^2\left(1-\rho(\bar{r}-\bar{r}'-v\tau)\right)+ik\kappa(\bar{r}-\bar{r}'-\bar{\Delta}).(\bar{r}+\bar{r}')-(r^2+r'^2)/W^2\right]$$

(6.30)

Assuming that $W^2 \gg \xi^2$ and $\varphi_0^2 \gg 1$, the following result is obtained

$$g^{(1)}(\bar{\Delta}, \tau) = \exp\left(-\frac{1}{2}\frac{k^2\kappa^2 W^2\left|\bar{\Delta} - \bar{v}\tau\right|^2 + v^2\tau^2/W^2}{1 + k^2\kappa^2 W^2\xi^2/2\varphi_0^2}\right) \tag{6.31}$$

When wave front curvature dominates, as in the Fresnel region, this reduces to

$$g^{(1)}(\bar{\Delta}, \tau) = \exp\left(-\xi^2\left|\bar{\Delta} - \bar{v}\tau\right|^2/\varphi_0^2\right) \tag{6.32}$$

The speckle pattern then has the same characteristic scale as the field at the screen and moves bodily across the detection plane with speed v. As pointed out previously, in the event of spatial averaging of the intensity of such a pattern by the detector, the fluctuation time will appear to be increased. Similarly, time averaging will appear to change the spatial characteristics of the radiation.

When the illuminating beam profile is sufficiently small, Equation 6.31 reduces to

$$g^{(1)}(\bar{\Delta}, \tau) = \exp\left[-\tfrac{1}{2}\left(k^2\kappa^2 W^2\Delta^2 + v^2\tau^2/W^2\right)\right] \tag{6.33}$$

The correlations therefore factorize and are said to be *cross-spectrally pure*. In this case the spatial scale characterizing the speckle pattern is determined by the size of the illuminated region, and the time taken for elements of the phase screen to cross the beam determines the rate of its evolution. This result is identical with that obtained for scattering by a uniformly moving particulate in the absence of curvature effects (see Section 4.7). Note that there is no interaction between space and time averages of the intensity pattern in this case.

6.12 Phase Statistics

In the preceding part of this chapter, attention has been devoted exclusively to fluctuations in the scattered *intensity* behind a smoothly varying Gaussian phase screen. There appear to be relatively few results in the open literature concerned with the evolution of the phase fluctuations. For the deep phase screen, where the total phase differences imposed by the screen are greatly in excess of 2π, the statistics of the phase modulo 2π at $z = 0$ will approach a uniform distribution regardless of the correlation function of the screen. The statistics will remain uniform for all distances z beyond the screen, so the distribution of the phase modulo 2π does not provide any information about the scattering process. It should be noted that, in the absence of other information, the phase at a point on the wave front can only be assigned a value modulo 2π, that is, between $-\pi$ and π or 0 and 2π, and so forth. This is evident in the cyclic nature of the dependence of the field E on the phase ϕ in the narrowband definition of Equation 1.39 and Equation 1.40.

In some circumstances, one may wish to construct an unwrapped phase from a series of measurements, each individually modulo 2π. For example, the phase φ of a phase screen characterized by the smoothly varying model of Equation 6.1 could be determined experimentally, to within a constant factor, by moving a heterodyne detector past the screen, in a direction perpendicular to z, and using a suitable phase-unwrapping algorithm (see [28] for an experiment very similar to this). However, problems arise in unwrapping the phase when there are strong intensity variations: it is not always possible to construct a single-valued two-dimensional unwrapped phase $\phi(\vec{r})$ from the field $E(\vec{r})$. The problem is closely related to the existence of points of zero intensity (nulls) in the field at which the phase becomes discontinuous [12]. For the deep random phase screen, these nulls start to develop in the focusing region and reach a maximum density in the Gaussian speckle limit [29,30].

Another phase-related quantity that is often of interest is the phase derivative, where the differentiation can be either with respect to one of the spatial transverse coordinates or with respect to time in the time-dependent case. The phase derivative can be unambiguously defined at every point of the scattered light field, although it will have singularities at the intensity nulls. If we consider the case of the static phase screen defined by Equation 6.1, the spatial phase derivative $\dot{\phi}$ will start at $z = 0$ as a Gaussian process with a correlation function found by applying Equation 2.34 to Equation 6.1. In the Gaussian speckle limit, at large z, its statistical properties will be given by Equation 2.38 and Equation 2.49. It can be shown more generally that the inverse cubic limit of the probability density (Equation 2.38) for large $\dot{\phi}$ arises from the behavior of the field in regions near the intensity nulls, and that a similar inverse cubic limit is expected for the probability density function of $\dot{\phi}$ in any situations where intensity nulls are present [28]. In such situations, the variance of $\dot{\phi}$ will be infinite because the kernel of the integral over the probability distribution is inversely proportional to $\dot{\phi}$ for large values, leading to a logarithmic divergence.

A statistical measure closely related to the phase derivative that always has a finite variance is the flux J defined by Equation 2.40. Since in many remote sensing systems it is the phase derivative that carries the information of interest, some recent calculations have sought to evaluate the statistical properties of J in the Fresnel region behind a deep random phase screen [31]. Using the steepest descents and modelling approach described in Section 6.4, the following analytical approximation was obtained for the corrugated Gaussian screen with Gaussian spectrum

$$\left\langle J^2 \right\rangle = \frac{\phi_0^2}{\xi^2} \left\{ erfc(q) + \frac{2q}{\sqrt{\pi}} \exp(-q^2) \left[\ln\left(\frac{8\gamma\phi_0^2}{3}\right) + \pi erfi(q) - \Re(q) - q^2 - 2 \right] \right\}$$

(6.34)

Here erfc (q) and erfi(q) are error functions (Ref. 4, Ch. 3) and

$$\Re(q) = 2\sqrt{\pi} \int_0^q dx \exp(x^2) erfc(x)$$

(6.35)

$$q = \frac{k\xi^2}{2z\phi_0\sqrt{6}}$$

A plot of the second moment of fluctuations in this quantity as a function of the inverse dimensionless propagation distance, q, is shown in Figure 6.16, comparison being made between theoretical results and numerical simulation. The plot bears some similarity to that for the scintillation index, with a peak reflecting the increase in amplitude fluctuations in the focusing region. However, the fluctuations at large propagation distances decrease to a value that is half that at the screen. Figure 6.17 shows probability densities of the phase derivative at different propagation distances obtained by numerical simulation, using the same parameters as in Figure 6.6 and Figure 6.7, and illustrates the development of the long inverse cubic tail expected for a Gaussian process at long propagation paths.

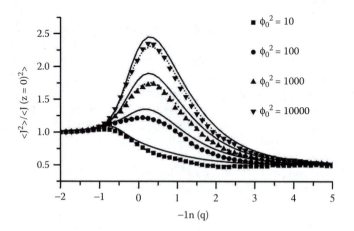

FIGURE 6.16

Normalized second moment of the intensity weighted phase derivative as a function of inverse dimensionless propagation distance q, comparison of the analytical result (Equation 6.34) with numerical simulation [31]. The solid lines are the result of Equation 6.34 and the dotted lines are a more accurate result that was calculated via a numerical integration.

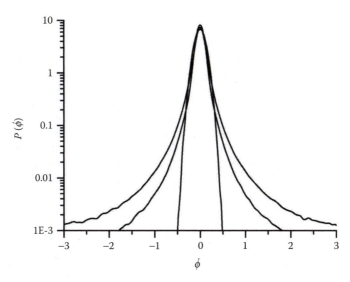

FIGURE 6.17
Simulation of scattering into the Fresnel region by a smoothly varying Gaussian phase screen, broadening of the probability density of the phase derivative with propagation distance.

References

1. Escamilla, H.M., and E.R. Mendez. Speckle Statistics from Gamma Distributed Phase Screens *J. Opt. Soc. Am.* A8 (1991): 1929–35.
2. Mercier, R.P. Diffraction by a Screen Causing Large Random Phase Fluctuations. *Proc. Camb. Phil. Soc.* A58 (1962): 382–400.
3. Salpeter, E.E. Interplanetary Scintillation I: Theory *Astrophys. J.* 147 (1967): 433–48.
4. Bramley, E.N., and M. Young. Diffraction by a Deeply Modulated Phase Screen. *Proc. IEE* 114 (1967): 553–56.
5. Singleton, D.G. Saturation and Focussing Effects in Radio-Star and Satellite Scintillations. *J. Atm. Terrest. Phys.* 32 (1970): 187–208.
6. Buckley, R. Diffraction by a Phase Screen with Very Large rms Phase Deviation. *Aust. J. Phys.* 24 (1971): 351–71.
7. Shishov, V.I. Strong Fluctuations of the Intensity of a Plane Wave Propagating in a Random Refracting Medium. *Sov. Phys. JETP* 34 (1972): 744–48.
8. Whale, H.A. Diffraction of a Plane Wave by a Random Phase Screen. *J. Atm. Terrest. Phys.* 35 (1973): 263–74.
9. Taylor, L.S. Diffraction Theory of Optical Scintillations due to Turbulent Layers. *J. Opt. Soc. Am.* 65 (1975): 78–84.
10. Jakeman, E., and J.G. McWhirter. Correlation Function Dependence of the Scintillation behind a Deep Random Phase Screen. *J. Phys. A: Math. Gen.* 10 (1977): 1599–1643.
11. Jakeman, E., and J.G. McWhirter. Non-Gaussian Scattering by a Random Phase Screen. *Appl. Phys.* B26 (1981): 125–31.

12. Nye, J.F. *Natural Focusing and Fine Structure of Light*. Bristol: IoP Publishing, 1999.

13. Parry, G., P.N. Pusey, E. Jakeman, and J.G. McWhirter. Focussing by a Random Phase Screen. *Opt. Commun.* 22 (1977): 198–201.

14. Tatarskii, V.I. *Wave Propagation in a Turbulent Medium*. New York: McGraw-Hill, 1961.

15. Berry, M.V. Focusing and Twinkling: Critical Exponents from Catastrophes in Non-Gaussian Random Short Waves. *J. Phys. A: Math. Gen.* 10 (1977): 2061–81.

16. Walker, J.G., M.V. Berry, and C. Upstill. Measurement of Twinkling Exponents of Light Focused by Randomly Rippling Water. *Optica Acta* 30 (1983): 1001–10.

17. Parry, G., P.N. Pusey, E. Jakeman, and J.G. McWhirter. The Statistical and Correlation Properties of Light Scattered by a Random Phase Screen. In *Coherence and Quantum Optics IV*. Eds. L. Mandel and E. Wolf . New York: Plenum, 1978, 351–61.

18. Booker, H.G., J.A. Ratcliffe, and D.H. Shinn. Diffraction from an Irregular Screen with Applications to Ionospheric Problems. *Phil. Trans. Roy. Soc.* 242 (1950): 579–609.

19. Hewish, A. The Diffraction of Radio Waves in Passing through a Phase-Changing Ionosphere. *Proc. Roy. Soc.* A209 (1951): 81–96.

20. Little, L.T., and A. Hewish. Interplanetary Scintillation and its Relation to the Angular Structure of Radio Sources. *Mon. Not. Roy. Astron. Soc.* 134(1966): 221–37.

21. Jakeman, E., R.C. Klewe, P.H. Richards, and J.G. Walker. Application of Non-Gaussian Light Scattering to Measurements on a Propane Flame. *J. Phys. D: Appl. Phys.* 17 (1984): 1941–52.

22. Walker, J.G., and E. Jakeman. Non-Gaussian Light Scattering by a Ruffled Water Surface. *Optica Acta* 29 (1982): 313–24.

23. Pusey, P.N., and E. Jakeman. Non-Gaussian Fluctuations in Electromagnetic Radiation Scattered by a Random Phase Screen, II: Applications to Dynamic Scattering in a Liquid Crystal. *J. Phys. A: Math. Gen.* 8 (1975): 392–410.

24. Walker, J.G., and E. Jakeman. Observation of Sub-Fractal Behaviour in a Light Scattering System. *Optica Acta* 31(1984): 1185–96.

25. Ward, K.D., and S.Watts. Radar Sea Clutter. In *Microwave Journal* June (1985): 109–21.

26. Jakeman, E., and K.D. Ridley. A Signal Processing Analogue of Phase Screen Scattering. *J. Opt. Soc. Am.* A15 (1998): 1149–59.

27. Jakeman, E., and P.N. Pusey. Significance of K-Distributions in Scattering Experiments. *Phys. Rev. Letts.* 40 (1978): 546–50.

28. Ridley, K.D., S.M. Watson, E. Jakeman, and M. Harris. Heterodyne Measurements of Laser Light Scattering by a Turbulent Phase Screen. *Applied Optics* 41 (2002): 532–42.

29. Baranova, N.B., B.Y. Zeldovich, A.V. Mamaev, and V.V. Shkunov. Dislocation Density on a Wave Front of a Speckle-Structure Light Field. *Sov. Phys. JETP*, 56, (1982): 983.

30. Ridley, K.D. Limits to Phase–Only Correction of Scintillated Laser Beams. *Opt. Comm.* 144 (1997): 299–305.

31. Jakeman, E., S.M. Watson, and K.D. Ridley. Intensity-Weighted Phase Derivative Statistics. *J. Opt. Soc. Am.* 18 (2001): 2121–31.

7

Scattering by Fractal Phase Screens

7.1 Introduction

In the previous chapter, attention was focused on smoothly varying, single scale, Gaussian phase screens with correlation functions of the type of Equation 6.1 that could be expanded in the even-powered series (Equation 6.2). This chapter examines the statistical properties predicted for radiation scattered by Gaussian phase screens characterized by power-law spectra. These can be used to model media that are characterized by a wide range of length scales.

Multiscale scattering models were developed in the 1960s by physicists interested in wave propagation through inhomogeneous media such as the atmosphere [1]. However, it was not until the concepts of fractal geometry [2] were brought to bear on the problem that the physical interpretation and mathematical implications of hierarchical phase screen models with power-law spectra were fully clarified. Authors engaged in many areas of physics have contributed to the subject, with a range of publications devoted to the effects of propagation through screens and extended media characterized by power-law spectra [3–15]. It is well known that turbulent fluid motion leads to velocity structures with a wide range of sizes, and that turbulent mixing of fluids parameterized by passive scalars such as temperature or humidity gives rise to density and therefore refractive index fluctuations reflecting this multiscale behavior. Variations in refractive index in turn introduce variations of phase into a propagating wave that eventually leads to changes in amplitude, whether propagation takes place through an extended region of turbulence or, as in the present model, through a concentrated layer that can be treated as a phase screen.

The standard Kolmogorov spectral model for turbulence expresses the correlation properties of velocity components and passive scalars such as temperature in terms of the *structure function* or variance of increments. The relevant statistic in the present phase screen scattering context is the spatial phase structure function and in contrast to Equation 6.1 the generic multiscale model for this may be expressed in the following form

$$D(\bar{\Delta}) = \left\langle \left[\varphi(\bar{r} + \bar{\Delta}) - \varphi(\bar{r}) \right]^2 \right\rangle = 2\varphi_0^2 \left[1 - \rho(\bar{\Delta}) \right] = k^2 L^{2-\alpha} \left| \Delta \right|^\alpha \quad 0 < \alpha < 2 \quad (7.1)$$

Here, the parameter L is called the *topothesy* [16] and is the length over which the change in optical path and separation are equal, that is, the average incremental slope of the wave front is unity. The index $\alpha = 5/3$ for the case of Kolmogorov turbulence, but consideration will not be confined to this special case since it is only one example of a class of models that have many features in common. It is clear from Equation 7.1 that the magnitude of the wave-front distortions increases with separation without limit and the phase is in fact not a stationary process in this model. However, the phase increments that are relevant to the scattering problem *are* stationary and the *effective* rms slope of the wave front, $\sqrt{D}/\Delta$, decreases with separation being proportional to $\Delta^{\alpha/2 - 1}$. Notice that the true slope of the wave front, defined as the first derivative of the phase, does not exist for the defined range of α and the relationship of Equation 2.34 between the correlation function of a variable and that of its derivative is not valid for variables with the power-law structure function of Equation 7.1. These are *continuous but not differentiable*.

In real life, the dependence (Equation 7.1) of D on separation might be observed in experimental data over several decades, but eventually the phase at two widely separated points will become uncorrelated. The range at which this occurs is usually called the *outer scale* l_0 of the fluctuations, and beyond this point the structure function will saturate at the value $2\varphi_0^2$ as the correlation function approaches zero. From Equation 7.1 it is apparent that the outer scale and mean square phase variation are related by $l_0 \approx \left(2\varphi_0^2 / k^2 L^{2-\alpha} \right)^{1/\alpha}$. It might be anticipated that the presence of an outer scale will be a *sufficient* condition for the statistics of waves scattered by a fractal phase screen to converge to Gaussian statistics when a large enough area of the screen contributes to the scattered intensity pattern. However, it turns out that this is not always a *necessary* condition.

In practice, the model of Equation 7.1 also becomes invalid at sufficiently *small* separations. For example, the temperature of the atmosphere will certainly change smoothly over distances smaller than the heat diffusion length, whatever the smallest scale of its turbulent motion. The first separation-dependent term in an expansion of the correlation function will then be quadratic as in the smoothly varying model discussed in Chapter 6. The incident wave will only sense this *inner scale* smoothing, however, if the wave front distortions over an inner length scale are of the order of a wavelength or more. According to Equation 7.1 this occurs if $l_i > \left(\lambda^2 L^{\alpha-2} \right)^{1/\alpha}$.

In light of the discussion in earlier chapters, only when the phase is a Gaussian process does Equation 7.1 complete the description of its statistical properties, and so this model will be adopted in the present chapter. The scattered wave front is then a *Gaussian random fractal* and the index α can be related to the Hausdorff-Besicovich fractal dimension H. The mathematical niceties of such characterizations are not of concern here but for completeness it may be noted that for a corrugated wave front (or for a section through an isotropic one) it can be shown that [10]

$$\alpha = 2(2 - H) \tag{7.2}$$

Thus a small index corresponds to a very rough, space filling wave front with $H \approx 2$, while when $\alpha = 2$ (a case that is actually excluded from the range) Equation 7.1 is just the first term in the expansion of Equation 6.2 corresponding to the more familiar differentiable one-dimensional curve with $H = 1$. Note that it is necessary to treat the asymptotic case $\alpha \to 2$, generally referred to as a *marginal* fractal [10], separately in the scattering context. In the special case $\alpha = 1$ ($H = 3/2$), realizations of the scattered wave front may be generated by a random walk of independent steps, giving rise to the name *Brownian fractal*. Cases for which $\alpha > 1$ can be generated by random walks of correlated steps while when $\alpha < 1$ realizations of the wave front can be generated by a random walk of anticorrelated steps (see Chapter 14 on numerical simulation).

Figure 7.1 shows simulations of corrugated fractal wave fronts together with expanded portions of the curves as insets. It is clear that the general character of the phase variations is preserved under magnification. This property is embodied in Equation 7.1 for if the phase increment on the left-hand side is scaled by a factor q, say, and the space increment on the right-hand side is scaled by $q^{2/\alpha}$, then the form of the structure function is preserved. The wave front is said to be *self-affine* rather than *self-similar* since the phase and the space dimension (effectively the strength and extent of the wave-front distortions) require *different* scaling for the structure function to remain invariant.

7.2 Scattering into the Fresnel Region by a Fractal Phase Screen

According to Equation 2.34, the correlation function of the derivative of the phase is proportional to the double derivative of Equation 7.1. However, this takes unphysical values when α is in the indicated range as a consequence of the fact that an object with the structure function of Equation 7.1 is

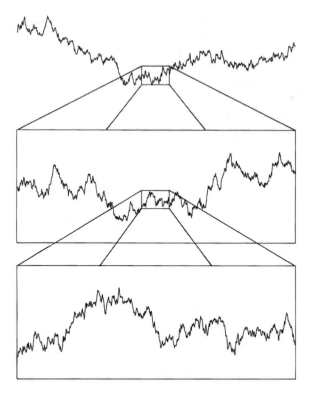

FIGURE 7.1
Corrugated fractal wave front with insets showing magnified portions.

continuous but not differentiable. This is an important observation since rays
or normals to the scattered wave front are not meaningful concepts if the
slope of the wave front is not defined, and it implies that fractal phase screens
will not generate geometrical optics effects. Since it was argued in the previous chapter that focusing and caustics are a principal cause of enhanced
non-Gaussian fluctuations in phase screen scattering, it might be expected
that radiation scattered by fractal screens will not display such effects. It is
indeed not difficult to demonstrate this for the case of a corrugated Brownian
fractal screen. The Fresnel region scintillation index can be calculated exactly
for this model from the one-dimensional version of Equation 7.1 with $\alpha = 1$
and the general Gaussian screen Equation 6.4 and may be expressed in terms
of Fresnel integrals [9]

$$S^2 = 1 - 2\left\{\left[\tfrac{1}{2} - C\left(\sqrt{k^3 L^2 z / 2\pi}\right)\right]^2 + \left[\tfrac{1}{2} - S\left(\sqrt{k^3 L^2 z / 2\pi}\right)\right]^2\right\} \qquad (7.3)$$

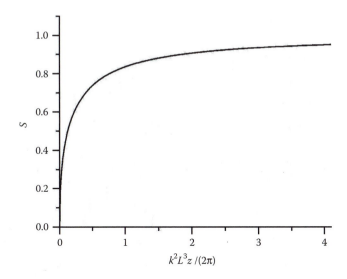

FIGURE 7.2
Scintillation index in the Fresnel region behind a Brownian fractal phase screen.

A plot of this quantity as a function of propagation distance in Figure 7.2 clearly shows how the scintillation index increases monotonically from zero at the screen to unity at large distances with no evidence of a focusing peak. It has been shown that the Rice distribution (Equation 3.5) provides a reasonable model for the single point intensity statistics at all propagation distances in this case [17]. In fact, calculations show that weak peaks are predicted for the higher allowed values of α, particularly in the case of two-dimensional screens, but these are far less pronounced than in the case of smoothly varying screens that generate geometrical optics effects. Infrared scattering from laboratory surfaces with fractal height spectra (Figure 7.3) have confirmed these predictions [18]. The observed non-Gaussian intensity patterns shown in Figure 7.4 are qualitatively different from those generated by smoothly varying diffusers (see Figure 6.2) and have manifestly lower contrast [19,20].

Another noteworthy feature of the result of Equation 7.3 is that it depends only on a single parameter that combines wavelength, topothesy, and propagation distance. It is not difficult to show more generally from Equation 6.4 and Equation 7.1 that the scintillation index in the Fresnel region of a Gaussian fractal screen is only a function of the single parameter $k^{4-\alpha}L^{4-2\alpha}z^{\alpha}$ for any value of α within the allowed range. Thus a particular value of S may be achieved at different distances for different wavelengths according to a simple scaling relationship, and it is impossible to distinguish the short wave limit from that of long propagation path, for example.

The intensity fluctuation spectrum in the Fresnel region of a Brownian fractal phase screen can also be obtained in closed form [9]

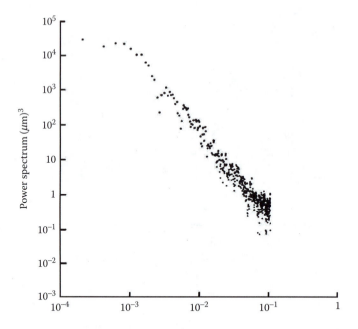

FIGURE 7.3
Structure function of an artificially constructed fractal surface [19].

$$\tilde{g}^{(2)}(K) = \pi\delta(K) + \frac{a}{a^2 + K^2}\left\{1 - \frac{\exp(-bK/a)}{K}\left[a\sin\left(bK^2/a^2\right) + K\cos\left(bK^2/a^2\right)\right]\right\}$$

(7.4)

Here the parameters are given by $a = k^2L$, $b = k^3L^2z$. The corresponding autocorrelation functions may be reduced to the expressions

$$g^{(1)}(\chi) = \exp(-a\chi)$$

$$g^{(2)}(\chi) = 1 + \left|g^{(1)}(\chi)\right|^2 - 2\int_0^\infty dx\cos a\chi x\frac{\exp(-bx)}{x(1+x^2)}\left[\sin bx + x\cos bx\right]$$

(7.5)

The result of Equation 7.5 is plotted as a function of the separation $a\chi$ for different values of the range parameter b in Figure 7.5. In the long-range limit, where b becomes very large, the integral in Equation 7.5 becomes small and the Siegert relation is recovered despite the absence of an outer scale in Equation 7.1. The parameter $1/a$ can be identified as the characteristic speckle size that will dominate the pattern in this asymptotic region. The fact that the Gaussian speckle limit is achieved without the presence of an outer scale suggests that different elements of the scattered wave front must

FIGURE 7.4
Scattered intensity pattern in the Fresnel region behind a fractal surface.

nevertheless act as independent scatterers. It is plausible that the size of these elements is equal to the distance over which the wave-front distortion changes by a wavelength. According to Equation 7.1 this is $1/k^2L$ and the results above are valid only if the outer scale is much larger than this size, which is equivalent to requiring that $\varphi_0^2 \gg 1$, that is, a deep phase screen. The width of the diffraction lobe at a distance z from such an element is of order $z/k \, (1/k^2L) = kLz$ and is precisely the spatial scale b/a appearing in the result of Equation 7.4. The frequency of the oscillations visible at smaller values of b is equal to the Fresnel zone size $\sqrt{z/k}$ that limits the region contributing to the pattern at short distances beyond the screen. However, at longer distances the diffraction lobes from different elements of the screen will overlap, and the effective illuminated area will be of order kLz. This occurs when $kLz > \sqrt{z/k}$, that is, when $b > 1$. The effective number of scattering centers will then be of order b. This interpretation is consistent with the speckle size $z/k.kLz = 1/a$, as found above and also with the observation that the first term in an expansion of the scintillation index or intensity correlation function in the asymptotic Gaussian regime is proportional to $1/b$ as expected from the random walk model (Equation 4.12). Note however

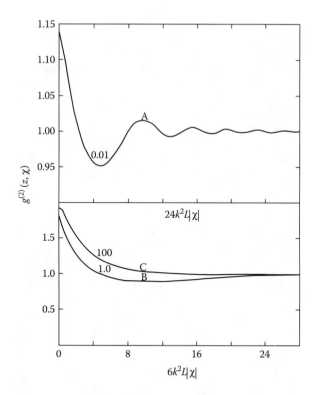

FIGURE 7.5
Spatial coherence function in the Fresnel region behind a Brownian fractal screen [9].

that, unlike the smoothly varying result Equation 6.16, these terms are negative, reflecting the absence of strong single scatterer fluctuations, that is, the final term in Equation 4.12.

The physical interpretation (above) of scattering by the Brownian fractal screen can be extended to other corrugated screens with values of α in the allowed range. Thus the size of the scatterers in the more general case is $\left(k^2 L^{2-\alpha}\right)^{-1/\alpha}$, and the corresponding diffraction scale is $(kL)^{2/\alpha-1}z$. When the propagation distance is sufficiently small, Fresnel zone oscillations of frequency $\sqrt{k/z}$ may be expected, but when $z > L(kL)^{1-4/\alpha}$ the diffraction lobes from different surface elements will begin to overlap. The contributing area of the screen will then increase as $(kL)^{2/\alpha-1}z$ and the speckle size will be $L(kL)^{-2/\alpha}$. It is not difficult to calculate the first-order coherence function from Equation 5.6 for a corrugated Gaussian fractal diffuser in order to check this

$$g^{(1)}(\chi) = \exp[-\tfrac{1}{2}D(\chi)] = \exp(-\tfrac{1}{2}k^2 L^{2-\alpha}|\chi|^{\alpha}) \qquad (7.6)$$

The $1/e$ point of this function is of the same order as the speckle size predicted by the heuristic arguments above. However, the scatterers are not independent except in the case $\alpha = 1$. This can be demonstrated by calculating the correlation of small increments

$$\left\langle \left[\varphi(0) - \varphi(\chi) \right]\left[\varphi(x) - \varphi(x + \chi) \right] \right\rangle \approx \tfrac{1}{2}\alpha(\alpha - 1)L^{2-\alpha}x^{\alpha-2}\chi^2 \qquad (7.7)$$

This result predicts correlation when $\alpha > 1$ and anticorrelation when $\alpha < 1$. As a consequence of this, the *effective* number of independent scatterers does not scale linearly with propagation distance but rather with $z^{2-\alpha}$. The deviation of the scintillation index from the Gaussian value of 2 is therefore asymptotically proportional to $z^{\alpha-2}$ [2,7] and as $\alpha \to 2$ the convergence becomes logarithmically slow [10].

7.3 Scattering into the Fraunhofer Region by a Fractal Phase Screen

Scattering into the Fraunhofer region by fractal screens has been the subject of a number of experimental and theoretical investigations, [9,19,21,22] mostly in connection with optical characterization of rough surfaces. An important signature of this kind of scattering object is the behavior of the distribution of scattered intensity with angle. The corrugated Brownian case provides an exactly solvable example that illustrates the point. Substituting model of Equation 7.1 with $\alpha = 1$ into the integral in Equation 6.26 obtains

$$\left\langle I \right\rangle \propto \left[\left(kL \right)^2 + \sin^2 \theta \right]^{-1} \qquad (7.8)$$

Comparison with the result of Equation 6.26 for a Gaussian spectral model shows that Equation 7.8 decreases very slowly with angle (Figure 7.6). The width of the distribution is governed by diffraction from the independent scattering elements identified above and not by the slope distribution of the scattered wave front as in the result of Equation 6.26. This characteristic is common to all members of the class of Equation 7.1, which give rise to scattered intensity distributions that decrease asymptotically like $(\sin \theta)^{-(1+\alpha)}$ in the corrugated case or $(\sin \theta)^{-(2+\alpha)}$ for the case of isotropic screens. Optical scattering experiments on artificially constructed fractal surfaces have confirmed this distinctive behavior [21]. It is interesting to note that the distributions of intensity with angle are members of the Levy stable class [23] that have assumed importance in connection with the phenomenon of self-organized criticality [24]

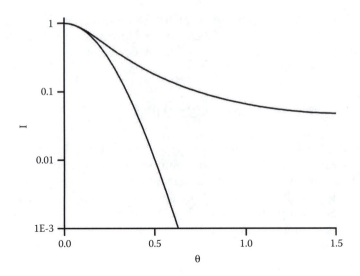

FIGURE 7.6
Scattered intensity as a function of angle in the Fraunhofer region, smoothly varying screen, (lower curve); Brownian fractal screen (upper curve).

$$\langle I(\theta)\rangle \propto p_\alpha\left(\left[kL\right]^{1-1/\alpha}\sin\theta\right)$$

where

$$\int_{-\infty}^{\infty} dx p_\alpha(x)\exp(i\lambda x) = \exp\left(-A\lambda^\alpha\right)$$

As in the Fresnel region, theoretical results for scattering by Gaussian fractals into the Fraunhofer region can generally be expressed in terms of products of parameters including the incident wavelength. Moreover, the absence of geometrical optics effects again precludes the development of large intensity fluctuations in Fraunhofer scattering configurations, even when the illuminated area is small. Calculations and experimental measurements show that as the illuminated area increases from being much smaller to being much larger than the effective scattering element size, the scintillation index in the forward-scattering direction increases from zero to unity either monotonically or with only a weak peak [19,25] (Figure 7.7). The dimensionless parameter governing the transition is $W^\alpha k^2/L^{\alpha-2}$ where W is the width of the illuminating beam. Although there is some experimental evidence for this behavior [19], few theoretical results are to be found in the literature.

As the scattering angle increases, there is a slow increase in the contrast of the pattern due to a reduction in the effective number of scattering centers

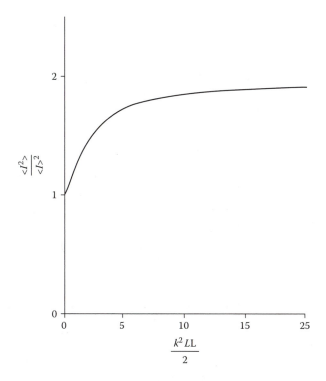

FIGURE 7.7
Scintillation index as a function of illuminated length L in the Fraunhofer region of a corrugated Brownian fractal phase screen [25].

[25] (Figure 7.8). However, this is a much weaker effect than in the case of smoothly varying screens since the contribution from each scatterer is more diffuse, being governed by diffraction rather than geometrical optics.

7.4 Subfractal Phase Screens

The results above apply provided the outer scale l_0 is much greater than the scattering elements on the wave front. It was conjectured earlier that these were of size $(k^2L^{2-\alpha})^{-1/\alpha}$, implying that in Equation 7.1 $\varphi_0^2 \gg 1$. This is in line with the deep phase screen assumption that has been made throughout. However, the presence of an inner scale over which the wave-front distortions or optical path differences are comparable to the wavelength of the incident radiation, profoundly changes the properties predicted for the scattered wave. It is possible to investigate this, while at the same time retaining the multiscale nature of the model, by smoothing a fractal wave front through integration over a short interval. Such an effect could be caused by

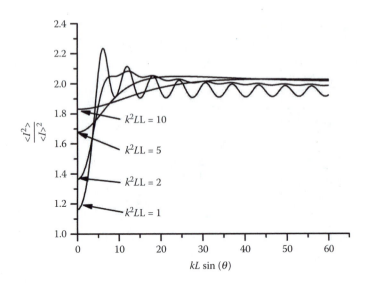

FIGURE 7.8
Scintillation index as a function of scattering angle in the Fraunhofer region behind a corrugated Brownian fractal phase screen for the values of k^2L L shown.

thermal diffusion, for example, in the case of refractive index fluctuations arising from temperature variations. The result is to generate a wave-front distortion with fractal slope or *subfractal.*

By definition, the slope of a subfractal wave front exists and therefore so does every normal or ray from this wave front. Thus it might be expected that such an object will generate geometrical optics effects. However, by construction, the slope of the scattered wave front is a fractal and so higher derivatives, including the curvature, are not defined. This means that although amplitude variations arise due to variations in the density of rays, there are no focusing points or caustics. A schematic illustration of this kind of phase screen scattering is shown in Figure 7.9 and may be contrasted with Figure 6.1. In light of the previous discussion, it is evident that the results of this kind of model phase screen will lie somewhere between those of the smooth single-scale model (Equation 6.1) and the fractal model (Equation 7.1). Since experimental data often seem to fall in this regime, the subfractal model is worthy of further investigation. As before, a joint Gaussian statistical model for the screen will be adopted so that in addition only the phase structure function is required to complete the model.

The phase structure function defining the subfractal or fractal slope model is [11,13]

$$D(\vec{r}) = \left(m_0^2 r^2 - |\vec{r}|^{\alpha+2} \right)\left(k^2 L^{-\alpha} \right)\left[(\alpha+1)(\alpha+2) \right]^{-1}, \quad for \quad |\vec{r}| \le l_0$$

$$= B \qquad\qquad\qquad\qquad\qquad\qquad for \quad |\vec{r}| > l_0$$

(7.9)

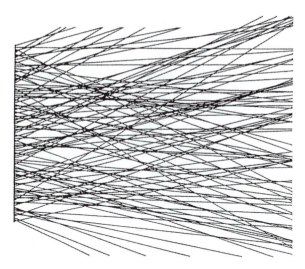

FIGURE 7.9
The pattern of rays behind a subfractal phase screen.

Here B is a constant and $0 < \alpha < 2$. In the case of a *section* through the wave front, Equation 7.9 would correspond to a slope structure function of the form

$$M(x) = \left\langle \left(m(0) - m(x) \right)^2 \right\rangle = \begin{cases} \left(|x|/L \right)^{\alpha}, & |x| \leq l_0 \\ 2m_0^2, & |x| > l_0 \end{cases} \tag{7.10}$$

where $km(x) = d\varphi(x)/dx$. An outer scale cutoff has been introduced to limit the mean square slope to the value

$$m_0^2 = \frac{1}{2}\left(\frac{l_0}{L} \right)^{\alpha} \tag{7.11}$$

L is the distance over which the change in the root mean square slope is unity. A number of theoretical results have been obtained for phase screen scattering using the model of Equation 7.9. [11,13,15,26] The behavior of the scintillation index and second-order coherence function $g^{(2)}$ is determined by the second power-law term, and they are finite in the limit of no outer scale, $l_0 \to \infty$. This regime will be examined first in some detail. Fresnel region plots of the scintillation index for various values of α are shown in Figure 7.10 as a function of the dimensionless propagation parameter

$$\beta = \left(\frac{z}{k} \right)\left(\frac{k^2}{2L^{\alpha}(\alpha + 1)(\alpha + 2)} \right)^{\frac{2}{(2+\alpha)}} \tag{7.12}$$

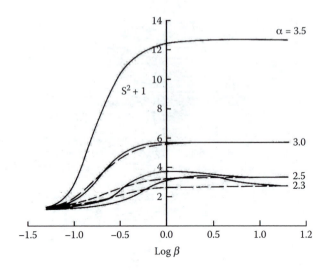

FIGURE 7.10
Scintillation index as a function of propagation parameter in the Fresnel region behind subfractal screens with no outer scale [15].

which characterizes the behavior in this limit. The graphs are obtained by evaluating Equation 6.4, using the model of Equation 7.9 with $l_0 \rightarrow \infty$ in the integrand. It is evident that the fluctuations saturate at large distances (or short wavelengths) at values that are greater than those expected for a Gaussian process. This kind of super-Gaussian saturated regime was predicted in the late 1970s, and the asymptotic values of the scintillation index can be calculated analytically. [11,13,15]

$$\lim_{\beta \to \infty} S^2 + 1 = \frac{4\sqrt{\alpha + 1}}{2 - \alpha} \quad \text{isotropic screen}$$

$$\lim_{\beta \to \infty} S^2 + 1 = \frac{4}{2 - \alpha} \quad \text{corrugated screen}$$

(7.13)

7.5 Ray Density Fluctuations beyond a Subfractal Phase Screen

The origin of the enhanced non-Gaussian fluctuations predicted by the result of Equation 7.13 is revealed through a study of the properties of the ray-density functional

$$R(\bar{r},z) = \frac{1}{z^2} \int\limits_{-\infty}^{\infty} d\bar{r}'\delta\left(\bar{m}(\bar{r}') - \frac{\bar{r}'-\bar{r}}{z} \right) \tag{7.14}$$

This describes the overlap of normals to the scattered wave front at a point $(\bar{r},z)$ beyond the screen. It may be shown that in the short wave limit the intensity moments can be expressed in term of the moments of this quantity, provided they are finite, as follows [13,15]

$$\left\langle I^N \right\rangle = N!\left\langle R^N \right\rangle \tag{7.15}$$

The mean value of the ray density is defined to be unity. It is not difficult to demonstrate that if m was smoothly varying then the variance and higher-order moments would be infinite and the formula is invalid. This reflects the infinite overlap of rays associated with the presence of caustics in the ray-density pattern. However, in the present case where m is a random fractal, the distribution of ray-density fluctuations has finite moments. In particular, assuming that the slope m is a Gaussian random fractal, it may readily be shown that in the absence of an outer scale the second moment of R is given by [13,15]

$$\left\langle R^2 \right\rangle = \frac{2\sqrt{\alpha+1}}{2-\alpha} \quad \text{isotropic screen}$$
$$= \frac{2}{2-\alpha} \quad \text{corrugated screen} \tag{7.16}$$

These values are one-half of those for the full diffraction treatment (Equation 7.13) in agreement with Equation 7.15. Note that the result of Equation 7.16 is independent of propagation distance. This is because if the slope of the wave front has no inner or outer scale, the statistics of the ray-density pattern do not evolve with distance apart from a simple length scaling. In the special case of a corrugated screen with Brownian fractal slope, $\alpha = 1$, analytical solutions for the entire complement of statistical properties can be obtained. Thus the correlation properties may be expressed as a sum over permutations of products of the first-order correlation function [13]

$$\left\langle R(y_1)R(y_2)....R(y_N)\right\rangle = \sum_n \prod_{j=1}^{N} \exp\left(\frac{-L}{z^2}\left|y_j - g_j^{(n)}\right| \right)$$

$$\left\langle R(0)R(\chi)\right\rangle = 1 + \exp\left(-\frac{2L|\chi|}{z^2} \right) \tag{7.17}$$

This is the well-known factorization theorem for the intensity correlation functions of a complex Gaussian process with Lorentzian spectrum and shows that the ray density is a gamma process of unit index (see, for example, Equation 3.19 and Equation 3.20 with $\alpha = 1$). Note the unusual dependence of Equation 7.17 on the propagation distance that implies that the structure in the pattern grows like the square of z rather than in the linear fashion that might have been expected from the result of Equation 6.16.

The moments of the ray-density distribution are given from Equation 7.17 by

$$\left\langle R^N \right\rangle = N!\tag{7.18}$$

Thus, for a corrugated *Brownian* subfractal screen, the distribution of intensity fluctuations in the asymptotic regime where Equation 7.15 is valid has moments $(N!)^2$ corresponding to the probability density

$$P(I) = 2K_0\left(2\sqrt{I}\right)\tag{7.19}$$

This is a member of the K-distribution class that was encountered in Chapters 3 and 4 (see, for example, Equation 3.25 with $\alpha = 1$, $b = 1$).

The results above strongly suggest that in the short wave limit, $\beta \to \infty$, the amplitude fluctuations of waves scattered by a Brownian subfractal phase screen can be interpreted as being due to the modulation of a Gaussian interference pattern by variations in the underlying ray density. This is a continuum analog of step-number fluctuations in the random walk model for particle scattering discussed in Chapter 4 (see Section 4.6). Calculation of the higher moments of the ray-density functional for other values of α (Figure 7.11) indicate that the distributions of fluctuations in R are also close to members of the class of gamma variates (Equation 3.13). Thus the corresponding intensity fluctuations will be approximately K distributed.

7.6 Coherence Properties of the Intensity Beyond a Subfractal Screen

We have already remarked that in the absence of a finite outer scale the ray-density fluctuations do not average out at large distances beyond the screen so that the intensity fluctuations are super-Gaussian in limit $\beta \to \infty$. Some further light is cast on the statistics in this limit by the form of the intensity correlation function. In the corrugated Brownian subfractal case, this takes the form [13]

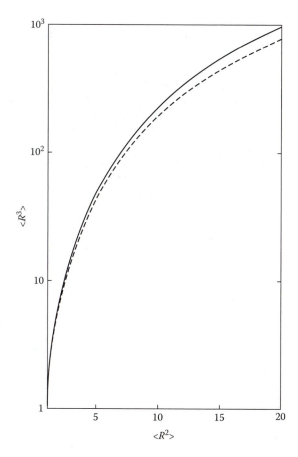

FIGURE 7.11
Comparison of the third moment of the distribution of ray-density fluctuations and the third
moment of a gamma distribution with the same variance and unit mean (shown dotted) [13].

$$\lim_{\beta \to \infty} \langle I(0)I(\chi) \rangle = 1 + \exp\left(-\frac{2L|\chi|}{z^2}\right) + \left(\frac{2\left\{1+\left[1+\left(2kz\chi/L\right)^2\right]^{1/2}\right\}}{1+\left(2kz\chi/L\right)^2}\right)^{1/2} \tag{7.20}$$

This reduces to Equation 7.13 with $\alpha = 1$ when $\chi = 0$. It does not obey the
Siegert factorization theorem (Equation 2.25) and cannot be expressed in
terms of the first order or field correlation function (Equation 5.7), which is
not defined for the model of Equation 7.10 in the absence of an outer scale.
The exponential term on the right-hand side of Equation 7.20 is a contribution due to ray-density fluctuations given by the result of Equation 7.17. The
final term is an interference term whose decay length reflects the weighting

imposed by wave-front geometry through the ray density and decreases with propagation distance. Thus, if it is assumed that the scale of the interference contribution to the intensity pattern (i.e., the speckle size) is z/kW_{eff}, where W_{eff} is the *effective* size of the scattering region contributing at the detector, then the last term in Equation 7.20 suggests that $W_{eff} \approx z^2/L$. This implies that the ray contributions will become independent at detector separations in excess of W_{eff}, a result that is consistent with the decay of the ray-density contribution to the pattern manifest in Equation 7.17 and the first term of Equation 7.20.

The unusual dependence of the ray-density pattern on distance implicit in the result of Equation 7.17 warrants further comment. It is not difficult to establish a more general scaling relationship for arbitrary values of α in the allowed range

$$\langle R(y_1)R(y_2)...R(y_N) \rangle = F\left(y_j \left[L^\alpha z^{-2} \right]^{1/(2-\alpha)} \right) \tag{7.21}$$

As a consequence, *all* features of the ray pattern *increase* like $z^{2/(2-\alpha)}$ with distance from the scattering screen, and this allows the arguments following Equation 7.20 to be generalized. The intensity correlation function can, in fact, always be expressed as the sum of two types of terms

$$\langle I(0)I(\chi) \rangle = F\left(\chi \left[L^\alpha z^{-2} \right]^{1/(2-\alpha)} \right) + G\left(\chi k \left[z/L \right]^{\alpha/(2-\alpha)} \right) \tag{7.22}$$
$$\lim \beta \to \infty$$

The first term is the direct ray-density contribution with the scaling shown in (7.21). The second term reflects the presence of interference effects with a speckle size $k^{-1}[L/z]^{\alpha/(2-\alpha)}$ that *decreases* with propagation distance in contrast to the smooth surface result of Equation 6.16. It is consistent, however, with the notion of an effective area contributing at the detector to give a speckle size z/kW_{eff} with $W_{eff} \approx (L^{-\alpha}z^2)^{1/(2-\alpha)}$. This interpretation is seen to be in agreement with the scaling of the ray-density pattern given by the first term in Equation 7.22.

7.7   Outer Scale Effects

The effect of a finite outer scale is to attenuate the ray-density fluctuations and thereby bring about convergence of the intensity statistics to Gaussian. Plots of the scintillation index versus the propagation parameter β are shown in Figure 7.12 for various values of the outer scale parameter $t = k^2 m_0^2 l_0^2$. The curves now decay towards the Gaussian value of two beyond a point where

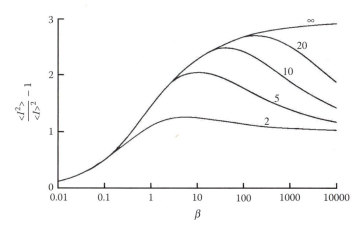

FIGURE 7.12
Scintillation index in the Fresnel region behind a subfractal phase screen as a function of normalized propagation distance for various values of *t* showing the effect of a finite outer scale [13].

the outer scale takes effect, giving rise to a more or less broad peak [13,15]. The second moment of the ray-density functional (Equation 7.14) may be evaluated for the corrugated case, for example, using Equation 7.10 and the result expressed in the form [13]

$$\frac{\langle R^2 \rangle}{\langle R \rangle^2} = 1 - \frac{1}{N} + \frac{1}{N} \langle R^2 \rangle \Big|_{l_0 \to \infty}$$

$$\text{with} \quad N = erf\left(\frac{1}{\sqrt{2}} \left[l_0 / W_{eff} \right]^{(2-\alpha)/2}\right)$$

(7.23)

This suggests that the attenuation of the ray-density fluctuations can be represented by the addition of $N(z)$ independent ray densities, each fluctuating according to the distribution obtained in the absence of an outer scale. If this occurs in the asymptotic regime where the ray density is approximately gamma distributed, then the attenuated fluctuations will also be gamma distributed since it is readily shown that the product of generating functions for independent gamma distributed variables is the generator for a gamma variable with a larger index (the mathematical property of *infinite divisibility*; see Chapter 12). As a consequence, the intensity fluctuations in the region beyond the peaks of the scintillation plots shown in Figure 7.12 will be approximately K-distributed. There is some evidence for this from experimental measurements of the statistical properties of light scattered by turbulent thermal plumes [27–29] (Figure 7.13).

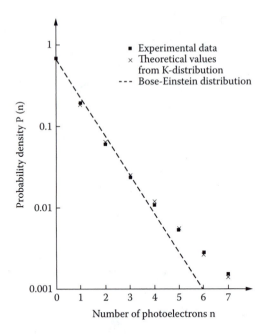

FIGURE 7.13
The photon-counting distribution of laser light scattered by a thermal plume compared with that expected in the case of K-distributed intensity fluctuations[28] (Equation 13.48).

For sufficiently large values of m_0 in the structure function of Equation 7.9 the field correlation function (Equation 5.7) is given for a corrugated sub-fractal screen by

$$g^{(1)}(\chi) = \exp\left(-\tfrac{1}{2}k^2 m_0^2 \chi^2\right) \tag{7.24}$$

while the second-order coherence function may be written

$$\lim_{k\to\infty}\langle I(0)I(\chi)\rangle = \langle R(0)R(\chi)\rangle +$$

$$+ 2\,\mathrm{Re}\left\{\frac{2erf\left(\gamma\sqrt{1+2ik\chi z/L}\right)}{\sqrt{1+2ik\chi z/L}} + \left|g^{(1)}(\chi)\right|^2 erfc\left(\gamma\sqrt{1+ik\chi z/L}\right)\right\} \tag{7.25}$$

where $\gamma = l_0/2zm_0$. This reduces to the Siegert relation for sufficiently large z. Thus in the large propagation distance limit, the presence of an outer scale ensures the statistics converge to Gaussian as expected.

7.8 Scattering into the Fraunhofer Region by a Subfractal Screen

Results for scattering by a Gaussian subfractal phase screen into the Fraunhofer region have also been derived using the model of Equation 7.9. [30,31] An approximate analytical formula for the scintillation index can be obtained from Equation 6.18 using a combination of steepest descents and function modelling as described in the previous chapter. For a Gaussian beam profile (Equation 4.3) the result takes the form [30]

$$S^2 + 1 = 2\exp(-s) + \frac{4s(s+t)^2}{s+2t} U \exp\left(\frac{k^2 W^2 st \sin^2 \theta}{(s+t)(s+2t)}\right)$$

(7.26)

$$U = \int_0^1 x dx \frac{\exp(-sx^2) - \exp(-r^2[2s + tx^\alpha])}{s + tx^\alpha}$$

Here $s = l_0^2/W^2$ and $t = m_0^2 k^2 l_0^2$. This result reduces to Equation 6.24 for the smoothly varying screen when $\alpha = 2$ and the integral can then be expressed in terms of tabulated exponential integrals. In that case, the first term on the right-hand side reflects the attenuation of interference effects as the illuminated area is reduced so as to contain few independently contributing areas of size $l_0(s \gg 1)$, while the second term expresses the balance between geometrical effects and aperture diffraction for a single-lens-like inhomogeneity of size l_0 and vanishes if the illuminated area is sufficiently large ($s \ll 1$). A similar interpretation is possible when $\alpha < 2$ except that inhomogeneities of size l_0 are now compound scatterers containing a significant though correlated substructure. The results of numerical evaluation of Equation 7.26 for two values of α are shown in Figure 7.14. The graphs are plotted as a function of $1/\sqrt{s}$ for various large values of t for which the analysis is valid, setting $\theta = 0$ for simplicity. As expected, the contrast increases from zero for small apertures, where aperture diffraction dominates, to a peak depending on the balance between geometrical and diffraction effects and finally decreases towards the Gaussian value of unity in the speckle regime where $s \ll 1$. The most interesting feature of the curves, however, is the emergence of a linear regime on the log-log plot at large values of t, particularly when α is small. Investigation of the integral in Equation 7.26 shows that this regime exists provided that $s \gg 1$ and

$$\left(s^{1+\alpha/2}/t\right)^{(2-\alpha)/\alpha} \ll 1$$

(7.27)

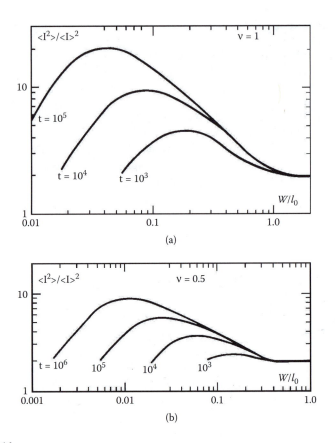

FIGURE 7.14
Scintillation index as a function of illuminated area in the Fraunhofer region behind a subfractal phase screen [30].

that is, for a sufficiently small aperture but sufficiently strong scattering. When these inequalities are satisfied, the integral can be evaluated approximately to give

$$S^2 + 1 \approx s^{\alpha/2} \exp\left(\sin^2 \theta / 2m_0^2\right) \tag{7.28}$$

On the other hand, when $s \ll 1$ an asymptotic regime similar to the result of Equation 6.15 for the smoothly varying phase screen is obtained

$$S^2 + 1 \approx 2(1-s) + \frac{2s}{(2-\alpha)} \exp\left(\sin^2 \theta / 2m_0^2\right)\left[1 - \Gamma\left(\frac{4}{2+\alpha}\right)\exp\left(\frac{\alpha-2}{\alpha+2}\ln t\right)\right]$$

$$\rightarrow 2(1-s) + \left(s/2\right)\exp\left(\sin^2 \theta / 2m_0^2\right)\ln t \quad as\ \alpha \rightarrow 2$$

$$\tag{7.29}$$

Since s can be interpreted as the inverse of the number of independent scattering centers, this indicates that the approach to Gaussian statistics is analogous to the random walk model (Equation 4.12) as expected.

The origin of the interesting second regime where $s \gg 1$ and Equation 7.27 is satisfied can be understood as follows. Geometrical effects begin to dominate aperture diffraction when the root mean square (rms) slope spread of the diffuser contained within the illuminated area is greater than the diffraction spread due to the aperture $\sqrt{M(W)} > (kW)^{-1}$, i.e., $s^{1+\alpha/2} < t$ (see Equation 7.10). Note that the peak contrast is achieved when $s^{1+\alpha/2} \approx t$, which, according to Equation 7.28 gives $S^2 + 1 \approx t^{\alpha/(2+\alpha)}$. On the other hand, interference effects can only be neglected if $s \gg 1$. These inequalities define the bounds of the region as in Equation 7.27 provided the index is not close to two. Within this region, scattering takes the form of a beam of rays from the illuminated area with characteristic angular divergence $M(W)$ and direction determined by the tilt of the larger scale sizes with an rms value of m_0. If the incident intensity is I_0, then the intensity per unit solid angle in the scattered beam falling on the detector is $I_0/M(W)$. The proportion of time spent by the beam on the detector, though, will be proportional to $M(W)/m_0^2$, so that a rough estimate of the moments of the distribution is given by

$$\frac{\langle I^n \rangle}{\langle I \rangle^n} \approx \frac{I_0^n}{M^n(W)} \cdot \frac{M(W)}{m_0^2} \bigg/ \left(\frac{I_0}{M(W)} \cdot \frac{M(W)}{m_0^2} \right)^{n-1} \approx \left(\frac{m_0^2}{M(W)} \right)^{n-1} \approx s^{\alpha(n-1)/2}$$

(7.30)

This result agrees with Equation 7.28 when $n = 2$. Note that the angle dependence in Equation 7.28 simply reflects the reduced probability of the beam falling on a detector placed off axis. Indeed a more general result can be deduced since the argument leading to Equation 7.30 does not depend on the phase being a Gaussian process. It is also worth pointing out that if the scattered beam gives intensity I_0 when it overlaps the detector but zero otherwise, then the nth intensity moment is just proportional to the mean given by

$$\langle I \rangle = I_0 \int_{S(W)} p(\bar{m} + \bar{u}\sin\theta)d\bar{m}$$

(7.31)

Here $p(\bar{m})$ is the slope distribution of the wave front, and $\bar{u}$ is a unit vector specifying the radial direction of the detector. In the region of interest, $\sqrt{M(W)}$ is much less than m_0, the width of the slope distribution, and so

$$\langle I^n \rangle \big/ \langle I \rangle^n \propto \left[2\pi M(W)p(\bar{u}\sin\theta) \right]^{1-n} = s^{\alpha(n-1)/2} \left[4\pi m_0^2 p(\bar{u}\sin\theta) \right]^{1-n}$$

(7.32)

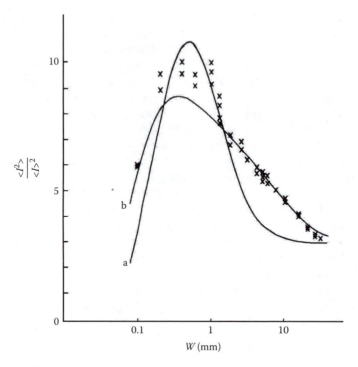

FIGURE 7.15
Second normalized moment of intensity fluctuations as a function of illuminated beam radius
in laser light scattered by a water-brine mixing layer compared to predictions of the subfractal
scattering model [32].

Data taken in experiments in which light was scattered through a layer of
turbulently mixing brine and water appear to exhibit both the power-law
dependence on illuminated area and the scaling of the moments of the
fluctuations at different angles predicted by Equation 7.32 [32] (Figure 7.15
and Figure 7.16).

7.9   Concluding Remarks

In conclusion, we have seen in this chapter that multiscale Gaussian phase
screen models generate intensity fluctuations that differ qualitatively from
those generated by single-scale smoothly varying models. On the whole,
fractal screens with no inner scale give rise to lower contrast patterns dominated by diffraction, and media that can be modelled in this way will impose
less severe limitations on system performance. Laboratory optical scattering
experiments with artificially constructed fractal surfaces show qualitative
agreement with the predictions of theory. There is also some numerical

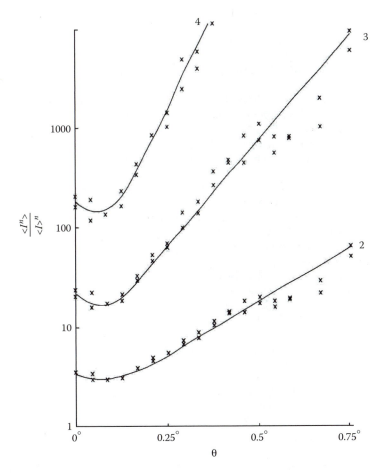

FIGURE 7.16
Angle dependence of higher moments of intensity fluctuations in light scattered by a water-brine mixing layer compared to predictions of the subfractal model [32].

evidence to suggest that the Rice process of Chapter 3 may provide an adequate model for the fluctuations in this case. The fractal slope model illustrates the effect of inner scale smoothing on multiscale phase screen scattering by introducing elementary geometrical optics effects. Its predictions are consistent with the notion of an interference pattern modulated by underlying ray-density fluctuations, analogous to the model of a random walk with step-number fluctuations described in Chapter 4. Experimental measurements of light scattering by a number of turbulent mixing layers suggest that this model provides a simple method of incorporating diffusion-mediated smoothing of the refractive index fluctuations that cause the observed scattering effects.

References

1. Tatarskii, V.I. *Wave Propagation in a Turbulent Medium.* New York: McGraw-Hill, 1961.
2. Mandelbrot, B.B. *The Fractal Geometry of Nature* San Francisco: Freeman, 1982.
3. Buckley, R. Diffraction by a Random Phase Screen with Very Large rms Phase Deviation. *Aust. J. Phys* 24 (1971): 351–71.
4. Shishov, V.I. Dependence of the Form of the Scintillation Spectrum on the Form of the Spectrum of Refractive Index Inhomogeneities. *Radiophys. Quant. Electron.* 17 (1974).
5. Gochelashvily, K.S., and V.I. Shishov. Saturation of Laser Radiance Fluctuations beyond a Turbulent Layer. *Opt. Quant. Electron.* 7 (1975): 524–36.
6. Rumsey, V.H. Scintillations due to a Concentrated Layer with a Power-law Turbulence Spectrum. *Radio Science* 10 (1975): 107–114.
7. Marians, M. Computed Scintillation Spectra for Strong Turbulence. *Radio Science* 10 (1975): 115–19.
8. Furuhama, Y. Covariance of Irradiance Fluctuations Propagating through a Thin Turbulent Slab. *Radio Science* 10 (1975): 1037–42.
9. Jakeman, E., and J.G. McWhirter. Correlation Function Dependence of the Scintillation behind a Deep Random Phase Screen. *J. Phys. A: Math. Gen.* 10 (1977): 1599–43.
10. Berry, M.V. Diffractals. *J. Phys. A: Math. Gen.* 12 (1979): 781–97.
11. Rino, C.L. A Power Law Phase Screen Model for Ionospheric Scintillation. *Radio Science* 14 (1979): 1135–45, 1147–55.
12. Uscinski, B.J., H.G. Booker, and M. Marians. Intensity Fluctuations due to a Deep Phase Screen with a Power Law Spectrum. *Proc. Roy. Soc. A* 374 (1981): 503–30.
13. Jakeman, E. Fresnel Scattering by a Corrugated Random Surface with Fractal Slope. *J. Opt. Soc. Am.* 72 (1982): 1034–41.
14. Uscinski, B.J., and C. Macaskill. Intensity Fluctuations due to a Deeply Modulated Phase Screen. *Atm. Terr. Phys.* 45 (1983): 595–605.
15. Jakeman, E., and J.H. Jefferson. Scintillation in the Fresnel Region behind a Subfractal Diffuser. *Optica Acta* 31 (1984): 853–65.
16. Sayles, R.S., and T.R. Thomas. Surface Topography as a Non-stationary Random Process. *Nature* 271 (1978): 431–34.
17. Roberts, D.L. The Probability Distribution of Intensity Fluctuations due to a Random Phase Screen with an Exponential Autocorrelation Function. In *Wave Propagation and Scattering.* Ed. B.J. Uscinski. Oxford: Clarendon Press, 1986, 130–53.
18. Jordan, D.L., R.C. Hollins, and E. Jakeman. Experimental Measurements of Non-Gaussian Scattering by a Fractal Diffuser. *Appl. Phys. B* 31 (1983): 179–86.
19. Jordan, D.L., E. Jakeman, and R.C. Hollins. Laser Scattering from Multi-scale Surfaces. *SPIE* 525 (1985): 147–53.
20. Jakeman, E. Optical Scattering Experiments. In *Wave Propagation and Scattering.* Ed. B.J. Uscinski. Oxford: Clarendon Press, 1986, 241–59.
21. Jordan, D.L., R.C. Hollins, and E. Jakeman. Measurement and Characterization of Multi-scale Rough Surfaces. *Wear* 109 (1986): 127–34.

22. Jordan, D.L., R.C. Hollins, A. Prewitt, and E. Jakeman. *Surface Topography* 1 (1988): 27–36.
23. Levy, P. *Theorie de L'Addition des Variables Aleatories.* Paris: Gauther-Villars, 1937.
24. Bac, P. *How Nature Works.* Oxford: OUP, 1997.
25. Walker, J.G., and E. Jakeman. Non-Gaussian Light Scattering from a Ruffled Water Surface. *Optica Acta* 29 (1982): 313–24.
26. Jakeman, E. Fraunhofer Scattering by a Subfractal Diffuser. *Optica Acta* 30 (1983): 1207–12.
27. Parry, G., P.N. Pusey, E. Jakeman, and J.G. McWhirter. Focussing by a Random Phase Screen. *Opt. Commun.* 22 (1977): 195–201.
28. Parry, G., P.N. Pusey, E. Jakeman, and J.G. McWhirter. The Statistical and Correlation Properties of Light Scattered by a Random Phase Screen. In *Coherence and Quantum Optics IV.* Eds. L. Mandel and E. Wolf. New York: Plenum, 1978, 351–61.
29. Ridley, K.D., S.M. Watson, E. Jakeman, and M. Harris. Heterodyne Measurements of Laser Light Scattering by a Turbulent Phase Screen. *Appl. Opt.* 41 (2002): 532–42.
30. Jakeman, E. Fraunhofer Scattering by a Sub Fractal Diffuser. *Optica Acta* 30 (1983): 1207–12.
31. Jakeman, E. Hierarchical Scattering Models. AGARD Conference Proceedings No. 419: *Scattering and Propagation in Random Media* (NATO 1988)
32. Walker, J.G., and E. Jakeman. Observation of Subfractal Behavior in a Light-Scattering System. *Optica Acta* 31 (1984): 1185–96.

8

Other Phase Screen Models

8.1 Introduction

In the previous two chapters the effect of the choice of phase spectrum on the fluctuation properties of waves scattered by joint Gaussian random phase screens was investigated. This choice of model for the phase variations was based in part on its well-known mathematical properties that facilitate analytical and numerical work. Also, in many practical situations of interest the exact nature of the phase fluctuation statistics is unknown and the Gaussian model is a plausible and convenient option that enables a qualitative understanding of the consequent properties of the scattered waves to be developed. In some situations the Gaussian model does indeed provide a fair representation of the first-order statistics, but the higher-order properties deviate from the joint Gaussian model. There is some evidence, for example, that the distribution of phase across a beam of light that has passed through a turbulent thermal plume may be gamma rather than Gaussian distributed [1]. Moreover, it is clear from common observation that there are many scattering objects introducing phase fluctuations into an incident wave that cannot be Gaussian at any level. This is particularly true of the rough interface between different media. For example, it is evident that neither the sea surface height nor its slope can generally be described by any kind of symmetric distribution since the crests of wind-driven waves are typically sharp and asymmetric while the troughs are shallow. In a completely different context, machined surfaces often consist of a series of grooves with rather irregular spacing, reflecting the imperfect nature of the cutting tool, so that the equivalent phase screen is effectively bimodal.

Thus, in phase screen scattering it is important to investigate not only the effect of different phase spectra on the properties of the scattered wave, but also how these are affected by the choice of the underlying stochastic model for the phase fluctuations. It is important to emphasize here that in the case of non-Gaussian statistics, the first-order spectrum of the fluctuations will *in general* be insufficient to completely define the statistical model. This is because the higher-order correlation properties cannot usually be expressed in terms of the first-order spectrum alone as in the factorization theorem governing the higher-order correlations of a Gaussian process (Section 2.5).

In the present chapter, properties of radiation scattered by two very different non-Gaussian statistical models for the random phase screen will be investigated. One model is effectively an extension of the joint Gaussian process, namely the generalized gamma process described in Chapter 3 [2–5]. The second is one in which the scattered wave front is a set of rectangular grooves of equal depth but irregular spacing so that its profile is a *random telegraph wave* [4,6,7]. The first of these models can again encompass both smoothly varying and multiscale behavior according to the choice of phase spectrum, while the random telegraph wave exhibits a variety of *fractal* behavior according to the choice of stochastic model for its zero crossings. Together they will provide some insight into the sensitivity of the scattered wave to the choice of basic statistical model rather than simply the spectral characteristics of the phase screen. It should be emphasized, however, that results for scattering by other models have occasionally figured in the literature [8–11].

8.2   Scattering by Smoothly Varying Gamma Distributed Phase Screens

The simplest scattering property that can be calculated for a gamma distributed phase screen is the angular distribution of scattered intensity in the Fraunhofer region. This may be obtained directly from the solution of Equation 5.16 and the joint generating function of Equation 3.17, taking the index to be ν rather than α so as to avoid confusion with spectral parameters. For a smoothly varying wave front having mean square slope $\langle m^2 \rangle$, the angular distribution of scattered intensity may be expressed in terms of modified Bessel functions of the second kind [2]

$$\langle I(\theta) \rangle \approx \left(\sin(\theta) \right)^{\nu - \frac{1}{2}} K_{\nu - 1/2} \left(\sin \theta \sqrt{2\nu / \langle m^2 \rangle} \right) \tag{8.1}$$

For large values of ν this reduces to Equation 6.26, since the generating function of Equation 3.17 approaches the joint Gaussian result of Equation 2.6 in this limit. However, for small values of the index, Equation 8.1 becomes singular in the forward direction and develops an exponential tail for large arguments of the Bessel function (Figure 8.1) rather than the Gaussian dependence on $\sin \theta$ exhibited by Equation 6.26. The predicted scattering distribution is therefore *qualitatively* different from the smoothly varying Gaussian model, and this may be attributed to the preponderance of small wave-front slopes when ν is small. As a consequence, in light scattering measurements

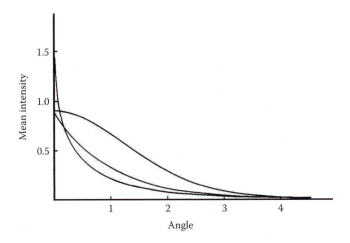

FIGURE 8.1

Mean intensity versus $\sqrt{2/\langle m^2 \rangle} \sin \theta$ scattered from smoothly varying gamma distributed phase screens with different values of v, lower (singular) curve $v = 0.5$, middle curve $v = 1$, upper curve $v = 9$ (Equation 8.1) [4].

on artificially constructed gamma distributed diffusers, [2] a significant component of unscattered radiation is observed in the specular direction. In the large aperture (Gaussian) limit, it is found that the transition from partially developed to fully developed speckle (see Section 4.8) with increasing surface roughness in this direction is very slow [2,5].

This "specular breakthrough" feature of the model is also manifest in the Fresnel region *weak* phase screen limit through results of the type in Equation 5.10. For example, using arguments analogous to those leading to Equation 5.15 together with result of Equation 3.17 with $s' = 0$, $s = i$, (recalling that the mean phase shift is finite in this model), the scintillation index at large propagation distances can be shown to be of the form

$$S^2 = 1 - \left(1 + \frac{\text{var } \varphi}{v} \right)^{-2v} \tag{8.2}$$

When the index is large, the final term approaches the Gaussian result $\exp(-2 \text{ var } \varphi)$ and decreases rapidly with mean square phase shift. However, when v is small, the convergence to unity with increasing phase fluctuation strength is very slow, and this can be interpreted as being due to the persistence of an unscattered contribution to the intensity also manifest in result of Equation 8.1.

One consequence of the tendency for specular breakthrough is that the analytical approximations that were used to evaluate the scintillation index and intensity coherence functions for the case of a smoothly varying Gaussian phase screen are less accurate. Nevertheless, they can be used to

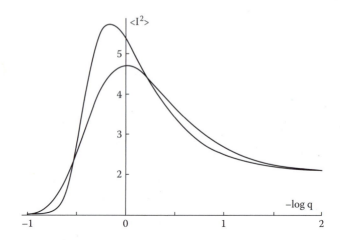

FIGURE 8.2
Mean square intensity as a function of the propagation parameter, $q = k\xi^2/z\sqrt{6\,\text{var}\,\varphi}$ in the Fresnel region behind a smoothly varying corrugated phase screen with variance 10^8, Gaussian model (narrower peak) and gamma distributed screen with $v = 1$ [3].

illustrate trends, provided var $\varphi \gg v$. The special case $v = 1$ can be evaluated using the combination of steepest descent approximation and function modelling described in Chapter 6 [3]

$$S^2 + 1 = 2e^{-q} + (q/2)\left[e^{-q}\ln\left(2\,\text{var}\,\varphi/3\right) + e^{-q}Ei(q) - e^q Ei(-q)\right] \qquad (8.3)$$

Here $q^2 = k^2\,\xi^4/6z^2\,\text{var}\,\varphi$ and Ei is the exponential integral. A more accurate numerical result for this case is plotted in Figure 8.2 for comparison with the Gaussian phase screen case $v \to \infty$. The qualitative behavior in this case is similar but the gamma distributed screen generates a broader, weaker focusing peak for the same phase variance.

8.3 Scattering by Fractal Gamma Distributed Phase Screens

In the case of hierarchical correlations, the average far field intensity is given by [4]

$$\left\langle I(\theta) \right\rangle \propto \int_0^\infty dx\, \frac{\cos\left(x\left[kL\right]^{1-2/\alpha}\sin\theta\right)}{\left(1 + |x|\alpha/v\right)^v} \qquad (8.4)$$

The behavior of this quantity is governed by diffraction from elements of the scattered wave front over which the displacement changes by a wavelength (see Sections 7.2 and 7.3), and the fall off at high angles with sin θ is essentially identical with that predicted for the Gaussian fractal screen. However, at small angles an integrable singularity is found for small v as in the case of a gamma distributed screen with smoothly varying spectrum, due in this case to the high probability of finding small changes in phase over a given distance (equivalent to small slopes for a smoothly varying screen).

When the phase is a gamma distributed fractal of unit index (i.e., Equation 7.1 with $\alpha = 1$), the scintillation index in the Fresnel region can be expressed without approximation in the following form [3]

$$S^2 + 1 = 2 - \frac{4}{\pi} \int_0^\infty \frac{dx}{x} \frac{\sin x^2}{\left(1 + z^{1/2}k^{3/2}Lx/2v\right)^{2v}} \tag{8.5}$$

This is plotted for various values of the index against the dimensionless propagation distance $p = \sqrt{kz}\, kL/2v$ in Figure 8.3. Again the behavior is qualitatively similar to that predicted for the Gaussian screen but the knee in the curve occurs at longer propagation paths for the gamma distributed case as the index is reduced.

No predictions are currently available for the intensity correlation functions of radiation scattered by gamma distributed phase screens, but some insight is provided by a study of ray density fluctuations in the case of the subfractal model (Equation 7.10). It is not difficult to demonstrate that the mean square ray density is identical to that obtained for the Gaussian case.

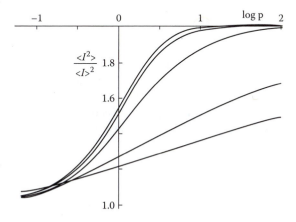

FIGURE 8.3
Scintillation index as a function of propagation parameter $p = \sqrt{kz}\, kL/2v$ in the Fresnel region behind a gamma distributed phase screen with linear structure function, lower to upper curves, $v = 0.05, 0.1, 0.5, 5, \infty$ [3].

Thus, for the corrugated screen we again have $\langle R^2 \rangle = 2/(2 - \alpha)$ for all values of v as in the Gaussian result (Equation 7.16). The correlation function of ray density fluctuations is given from Equation 7.14 in the corrugated case by [3]

$$\langle R(0)R(\chi) \rangle = \frac{1}{2\pi z} \int\limits_{-\infty}^{\infty} dx \int\limits_{-\infty}^{\infty} dy\, C(y,-y)\exp\left[iy\left(x - \chi\right)/z\right] \quad (8.6)$$

Here $C(a, b)$ is the characteristic function for the joint slope distribution of a gamma process. This can be derived by analytical continuation of the results for the sum of squares of Gaussian variables Equation 3.11. The calculation is straightforward but tedious and only the final result will be quoted here [4]

$$\begin{aligned}
C(a,b) = \Big\{ &1 - A^2 \left[\rho_0''(a^2 + b^2) + 2ab(\rho\rho'' + \rho'^2) \right] \\
&+ 2iA^3 ab(a - b)(\rho'\rho'' - \rho\rho'\rho_0'') \\
&+ A^4 a^2 b^2 \left[\left(\rho_0''^2 - \rho''^2 \right)(1 - \rho^2) + 2\rho'^2(\rho_0'' - \rho\rho'') + \rho'^4 \right] \Big\}^{-v}
\end{aligned} \quad (8.7)$$

In this expression, primes on the phase autocorrelation function ρ denote differentiation with respect to its argument, $A^2 = \text{var } \varphi/v$ and the subscript "0" denotes evaluation at zero separation. Substituting (8.7) into (8.6) leads to the following result for the "Brownian" case ($\alpha = 1$ in definition 7.10)

$$\langle R(0)R(\chi) \rangle = 1 + \frac{2}{\Gamma(v)} \left(\frac{\sqrt{v\chi L}}{2z} \right)^v K_v \left(\frac{\sqrt{v\chi L}}{z} \right)$$

The second term on the right-hand side has the same form as the K-distributions that were described in Chapter 3 (Equation 3.23 and Equation 3.24) and exhibit an exponential decay with the *square root* of χ at large separations. This may be contrasted with the Gaussian result (Equation 7.17) where the argument of the exponential factor is a linear function of χ (Figure 8.4).

8.4 Telegraph Wave Phase Screens

The second non-Gaussian phase screen model that will be considered in this chapter is the random telegraph wave, where the scattered wave front is a surface of rectangular grooves (Figure 8.5)

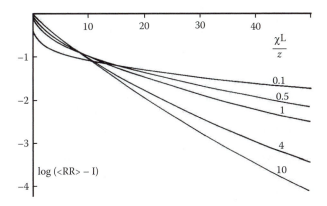

FIGURE 8.4
Spatial structure function of ray density fluctuations, gamma model with the values of v shown and linear slope structure function [3].

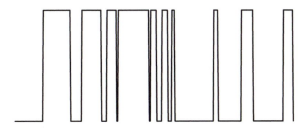

FIGURE 8.5
Telegraph wave phase screen.

$$\varphi(x) = khT(x) = \pm kh \tag{8.8}$$

From this definition it is evident that

$$\exp(i\varphi) = \cos kh + iT(x)\sin kh \tag{8.9}$$

Here h is the groove depth. Although the Huyghens-Fresnel approximation (5.2) is not strictly valid for wave fronts with discontinuities, investigation shows that it provides an adequate starting point for the prediction of diffraction beyond the near field [6].

From Equation 5.2 and Equation 5.16 the fields scattered into the Fresnel or Fraunhofer regions, respectively, by a telegraph wave phase screen can always be expressed as the coherent (vector) sum of constant and fluctuating components. It is clear from Equation 5.16 that in the Fraunhofer region the constant component is concentrated in the forward (specular) direction within the aperture diffraction lobe, while in the Fresnel region Equation 5.2 it is spatially uniform. Note that the constant component vanishes when the product kh is an odd multiple of $\pi/2$. Contributions to the field from the top and bottom faces of the grooves then interfere destructively while the fluctuating part of the field due to diffraction from the different groove widths survives. Only this part of the field is of interest here since the full solution to the scattering problem can easily be obtained by the coherent addition of an appropriate constant component. Thus the factor $\exp(i\varphi)$ in Equation 5.2 and Equation 5.16 is simply replaced by $T(x)$.

In order to evaluate statistical properties such as Equation 5.5 and Equation 5.8 of the radiation scattered by a phase distortion of the form in Equation 8.8, it is necessary to know the correlation properties of a random telegraph wave. These may be expressed in terms of the joint generating function for the numbers of crossings in different nonoverlapping intervals of space

$$Q\left(\{s_i\};\{x_i\}\right) = \sum_{\{n_i\}=0}^{\infty} P\left(\{n_i\};\{x_i\}\right) \prod_i \left(1-s_i\right)^{n_i} \tag{8.10}$$

In the case of a symmetric telegraph wave, for example we have for $x > 0$

$$\langle T(0)T(x)\rangle = \text{probability } (n \text{ even in } x) - \text{probability } (n \text{ odd in } x)$$

From (8.10)

$$\text{probability } (n \text{ even in } x) = \tfrac{1}{2}\left(Q(2;x)+Q(0;x)\right)$$

$$\text{probability } (n \text{ odd in } x) = \tfrac{1}{2}\left(Q(0;x)-Q(2;x)\right)$$

so that

$$\left\langle T(0)T\right\rangle(x) = Q(2;x) \tag{8.11}$$

An analogous but more lengthy argument leads for $y > x > 0$ to

$$\left\langle T(0)T(x)T(y)T(x+y)\right\rangle = Q(2,0,2;x,y-x,x) \tag{8.12}$$

For simplicity it will be assumed here that the crossings of the telegraph wave form a purely random series of events so that the *number* of crossings in a given time interval is Poisson distributed and governed by the well-known result $\langle(1-s)^N\rangle = Q(s;x) = \exp(-2sRx)$ where R is the average crossing rate. The generating functions of Equation 8.10 then factorize provided the intervals of space are ordered [12] and it may be shown that for $y > x > 0$

$$\langle T(0)T(x) \rangle = \exp(-2Rx) \tag{8.13}$$

$$\langle T(0)T(x)T(y)T(x+y) \rangle = \exp(-4Rx) \tag{8.14}$$

The first-order correlation function Equation 8.13 is evidently linear near the origin reflecting the nondifferentiable character of the telegraph wave but with an exponentially decaying memory or outer scale.

8.5   Scattering by Telegraph Wave Phase Screens

Using result of Equation 8.14 in the one-dimensional form of Equation 5.8 with $\Delta = 0$, it may readily be demonstrated that the scintillation index of the *fluctuating* contribution to the intensity pattern scattered into the Fresnel region is given by [13,3]

$$S^2 = 1 - 2 \left\{ \left[\tfrac{1}{2} - C\left(\sqrt{8zR^2/\pi k} \right) \right]^2 + \left[\tfrac{1}{2} - S\left(\sqrt{8zR^2/\pi k} \right) \right]^2 \right\} \tag{8.15}$$

This is identical in form to the result of Equation 7.3 obtained for scattering by the Gaussian (Brownian) fractal. Note, however, that although the dependence on propagation distance is identical, the scaling with wavelength is different. If the nonfluctuating contribution to the field is taken into account, the following formula is obtained

$$S^2 = \tfrac{1}{2} \left[1 - f(u) - g(u) \right] \sin^2 2kh + \left[1 - 2f^2(u) - 2g^2(u) \right] \sin^4 kh \tag{8.16}$$

Here f and g are auxiliary functions of the Fresnel integrals [14] and $u = 4R^2z/k$. The result of Equation 8.16 is plotted in Figure 8.6 and shows that there is a monotonic increase of the scintillation index with propagation distance for all phase variances. Saturation occurs at large distances where the distribution of intensity fluctuations is Ricean corresponding to partially developed Gaussian speckle.

The distribution of scattered intensity in the Fraunhofer region can be calculated from Equation 5.16 and is proportional to $[(2 R/k)^2 + \sin^2 \theta]^{-1}$. Comparison with the result of Equation 7.8 for scattering by a Gaussian (Brownian) fractal shows that the dependence on angle has the same functional form but again has different wavelength dependence.

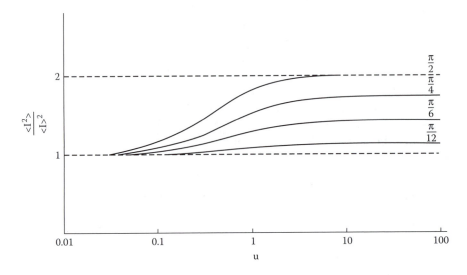

FIGURE 8.6
Scintillation index as a function of propagation parameter $u = 4R^2z/k$ in the Fresnel region behind a telegraph wave phase screen for the phase changes (kh) shown (Equation 8.16) [6].

A number of results exist for the scintillation index in the Fraunhofer region. The problem can be solved analytically if the illuminated region has sharp edges, that is, a "hard" aperture of length L, but the calculations are tedious and the mathematical results are in general fairly opaque [6]. Plots of the scintillation index against the average number of crossings within the aperture $\bar{N} = R\,L$ for this case is shown in Figure 8.7. If the constant term in Equation 8.9 is retained in the calculations, then, within the aperture diffraction lobe, the scintillation index reduces to zero in both large and small aperture limits due to specular breakthrough. It is clear from Equation 5.16 that the scattered field is real in the forward direction ($\theta = 0$) if the constant term in Equation 8.9 is neglected and in this direction it approaches a *real* Gaussian variable in the large aperture limit. As the aperture size is increased in this case, therefore, the scintillation index increases monotonically from zero to two. At other angles the field is a complex variable and the scintillation index increases monotonically from zero to unity as in the Gaussian fractal case.

Some further Fraunhofer region results can be obtained in the forward scattering direction for the telegraph wave model by noting that the scattered field is simply the telegraph wave filtered by an aperture function. Many results for this problem are to be found in the signal processing literature. In the case of a hard aperture, for example, the distribution of intensity fluctuations takes the form [6]

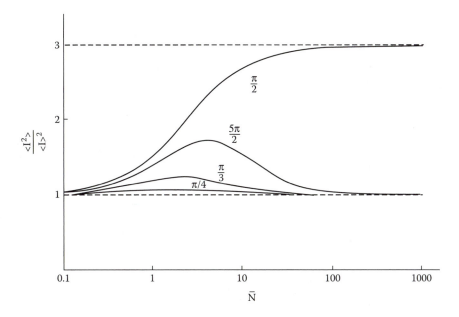

FIGURE 8.7
Scintillation index in the Fraunhofer region behind a telegraph wave phase screen as a function of illuminated length ($\bar{N} = R \times$ aperture size, L) [6].

$$P(I) = \frac{\exp(-N)}{2\sqrt{I}}$$

$$\left(N \left\{ I_0\left(N\sqrt{1-I}\right) + \frac{I_1\left(N\sqrt{1-I}\right)}{\sqrt{1-I}} \right\} + \delta\left(\sqrt{I}-1\right) + \delta\left(\sqrt{I}+1\right) \right) \quad 0 \le I \le 1 \tag{8.17}$$

Here $N = RL$ is the mean number of crossings in the aperture. The distribution is plotted in Figure 8.8. A more general result can be obtained for the asymmetric case where the mean value of the telegraph wave differs from zero [7].

On the other hand, for the case of an exponential aperture, $A = \exp(-\gamma x)$ and $\theta = 0$ in Equation 5.16 it may be shown that setting $\eta = 2R/\gamma$

$$P(I) = \frac{\Gamma\left(\frac{1}{2}+\eta\right)\left(1-I\right)^{\eta-1}}{\sqrt{\pi}\,\Gamma(\eta)} \frac{1}{2\sqrt{I}} \quad 0 \le I \le 1 \tag{8.18}$$

This is plotted for comparison with the hard aperture result in Figure 8.9. New results for the joint distribution of an exponentially filtered telegraph

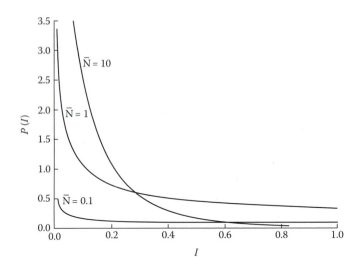

FIGURE 8.8
Telegraph wave phase screen, hard aperture, intensity fluctuation distribution in the forward scattering direction for the values of $\bar{N}$ shown [6].

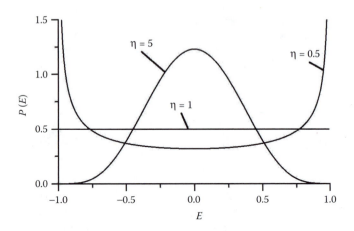

FIGURE 8.9
Telegraph wave phase screen, exponential aperture of width γ^{-1}, intensity fluctuation distribution in the forward scattering direction, $\eta = 2R/\gamma$ ($E = I$ in the text) [15].

wave have recently been derived and can be mapped onto the scattering problem [15].

The discussion above relates to scattering by a *symmetric* random telegraph wave. The theory has also been generalized to deal with the case of an asymmetric or *biased* telegraph wave for which $\langle T \rangle \neq 0$ [7]. In particular, a number of results have been derived for Fraunhofer scattering in the case of a hard aperture. It is found that the scintillation index in general exhibits

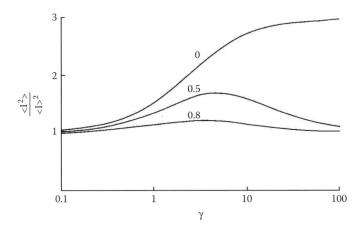

FIGURE 8.10
Scintillation index in the Fraunhofer region behind a biased telegraph wave phase screen as a function of normalized (hard) aperture size $\gamma = \bar{N}/(1 - \langle T \rangle^2)$ for the values of <T> shown [7].

a peak as a function of aperture size due to presence of the unscattered component (Figure 8.10). The peak is displaced by an amount that reflects the degree of bias (i.e., $\langle T \rangle$) and therefore provides a method for measuring this parameter independent of the absolute intensity.

8.6 Phase Statistics

In the previous sections of this chapter, only the amplitude statistics of the scattered wave have been discussed. In the case of smoothly varying gamma distributed phase screens, the distribution of phase fluctuations is effectively uniform in all regimes while the phase derivative statistics change fairly quickly from the exponentially decaying K-distribution (Equation 3.23, i.e., the distribution of the gradient of a gamma process) at the screen to the long-tailed students' t-distribution (Equation 2.38) in the regimes of Gaussian speckle. In the case of the fractal gamma distributed phase screen, the phase is again effectively uniform in all regimes while the phase derivative changes discontinuously from the screen, where it is not defined, to the students' t-distribution in the Gaussian speckle regime.

The situation is somewhat different in the case of scattering by a deep telegraph wave phase screen since, in general, there will be a constant, nonscattered contribution to the field in both Fresnel and forward Fraunhofer scattering configurations. In the Fresnel region, for example, the phase derivative distribution will evolve to that of a Rice process (Equation 3.8) at large propagation distances. In the Fraunhofer region, the behavior of the scattered field when the illuminated area encompasses many independent elements depends on the angle of view. Thus, in the forward direction the field is the vector sum of a constant component, X_0, and a *real* Gaussian process, Y, with $\tan \theta = Y/X_0$. The properties of the phase can therefore be calculated directly from those of Gaussian processes using a simple transform of variables (Section 1.3). For example, $\dot{\phi} = X_0 \dot{Y}/(X_0^2 + Y^2)$, where Y and its derivative are statistically independent Gaussian variables. At other angles the field is the sum of a constant component and a *complex* Gaussian, that is, a Rice process as in the Fresnel region.

8.7   Concluding Remarks

It is evident from the investigation given in this chapter that the angular distribution of intensity scattered by a deep phase screen into the Fraunhofer region is very sensitive to the statistical model adopted for the screen and is different for the models that have been studied. The higher-order intensity statistics are less sensitive to the underlying statistical model and appear to be largely determined by the spectral properties of the screen. Although fluctuation statistics such as the scintillation index appear to exhibit qualitatively similar behavior, important differences do occur in the rate of convergence to Gaussian statistics at large propagation distances in the Fresnel region and for large illuminated area in the Fraunhofer region. The transition from partially to fully developed speckle with increasing phase variance is also different for the different models.

In retrospect, it is not surprising that telegraph wave phase screen scattering is similar in some respects to scattering by a Gaussian Brownian fractal phase screen; neither model is differentiable, both being characterized by structure functions that are linear near zero separation, and both therefore generate only diffraction and interference effects associated with independent elements of the scattered wave front. Nevertheless, the role of their statistical properties as opposed to their spectral properties in governing the behavior of the scattered field is not entirely clear. Some light is shed on this problem by adopting a telegraph wave model in which the crossings are not random but generated by individuals leaving a population governed by a simple birth-death-immigration process [16]. The crossings will then be bunched with a correlation length ξ_c. The generating function for the number of events counted in an interval X is known exactly for this problem. For present purposes it is sufficient to take the limit when the number of crossings occurring in ξ_c is large, but where $X \ll \xi_c$, then

$$\langle T(0)T(X) \rangle \approx \left(1 + 2RX/v\right)^{-v} \qquad (8.19)$$

It is readily shown that this correlation function will give the same angular distribution of intensity as a gamma distributed phase screen with a linear structure function. Now in the high-density limit the birth-death-immigration process approaches the continuous gamma process with negative exponential correlation function, while the Poisson distribution tends to become Gaussian. Thus it appears that the crossing statistics of the telegraph wave model can be used to simulate the effects of the amplitude statistics of other models that are characterized by linear structure functions.

In conclusion, it has been demonstrated in this chapter that phase screen scattering is sensitive to the choice of statistical model as well as to the choice of phase spectrum. The distribution of intensity with angle, in particular, is qualitatively different for the models that have been investigated. It is therefore important to match the mathematical model to the underlying physics in the execution of any numerical simulations.

References

1. Schatzel, K. Interferometric Analyisis of Deep Random Phase Screens. *Opt. Lett.* 5 (1980): 389–91.
2. Kim, M.J., E.R. Mendez, and K.A. O'Donnell. Scattering from Gamma-Distributed Surfaces. *J. Mod. Opt.* 34 (1987): 1107–19.
3. Jakeman, E. Scattering by Gamma Distributed Phase Screens. *Waves in Random Media* 2 (1991): 153–67.
4. Jakeman, E. Non-Gaussian Statistical Models for Scattering Calculations. *Waves in Random Media* 3 (1991): S109–19.
5. Escamilla, H.M., and E.R. Mendez. Speckle Statistics from Gamma-Distributed Random Phase Screens. *J. Opt. Soc. Am.* A8 (1991): 1929–35.
6. Jakeman, E., and B.J. Hoenders. Scattering by a Random Surface of Rectangular Grooves. *Optica Acta* 29 (1982): 1587–98.
7. Jakeman, E., and E. Renshaw. Correlated Random Walk Model for Scattering. *J Opt. Soc. Am.* A4 (1987): 1206–12.
8. Kivelson, M.G., and S.A. Moszkowski. Reflection of Electromagnetic Waves from a Rough Surface. *J. Appl. Phys.* 36 (1963): 3609–12.
9. Beckmann, P. Scattering by Non-Gaussian Surfaces. *IEEE Trans. Antenn. Propag.* AP-21 (1973): 169–75.
10. Berry, M.V. The Statistical Properties of Echoes Diffracted from Rough Surfaces. *Phil. Trans. Roy. Soc.* A273 (1973): 611–58.
11. Jaggard, D.L., and Y. Kim. Diffraction by Band-Limited Fractal Screens. *J. Opt. Soc. Am.* A4 (1987): 1055–62.
12. Renshaw, E., and R. Henderson. The Correlated Random Walk. *J. Appl. Prob.* 18 (1981): 403–14.
13. Jakeman, E., and J.G. McWhirter. Correlation Function Dependence of the Scintillation behind a Deep Random Phase Screen. *J. Phys. A Math. Gen.* 10 (1977): 1599–1643.
14. Abramowitz, M., and I.A. Stegun. *Handbook of Mathematical Functions.* New York: Dover, 1971, 300.

15. Jakeman, E., and K.D. Ridley. Statistics of a Filtered Telegraph Signal. *J. Phys. A: Math. Gen.* 32 (1999): 8803–21.

16. Jakeman, E. A Simple Multi-Scale Scattering Model. *Optica Acta* 28 (1981): 435–51.

9

Propagation through Extended Inhomogeneous Media

9.1 Introduction

The previous four chapters discuss the scattering of waves by phase screens. The phase screen describes a scattering layer which is sufficiently short that its effect can be approximated by a change in the phase of the incident wave only; intensity fluctuations develop during propagation through free space after the screen. Thus it cannot be an accurate model for propagation through an extended medium when intensity fluctuations develop within the region of inhomogeneity. Nevertheless, the qualitative features of the scattering can be similar.

Propagation through extended inhomogeneous media has been the subject of entire books, particularly in the case of propagation of electromagnetic waves through the atmosphere [1–4], and will only be touched on here. In this chapter, some of the simpler approaches to the analysis of fluctuations will be discussed, making use of the phase screen methodology developed in previous chapters. The representation of a section of extended medium by a phase screen, or by multiple phase screens, will be described, particularly in relation to power-law media, and some of the probability distributions used to describe intensity fluctuations will be reviewed.

In addition to optical propagation through the atmosphere, this topic is relevant to propagation of sound waves through the ocean.

9.2 Single Phase Screen Approximation

The propagation of waves though a region of weak inhomogeneity is described by the paraxial wave equation with the inclusion of a varying refractive index n

$$\frac{\partial^2 E}{\partial x^2} + \frac{\partial^2 E}{\partial y^2} - 2ik\frac{\partial E}{\partial z} + 2nk^2 E = 0 \tag{9.1}$$

This is referred to as the parabolic approximation [5]. If the incident wave is plane the second derivatives will be initially zero. There will be some distance for which these derivatives can be neglected, a distance that will be longer when n is weaker or its spatial variations slower. Within this region the solution of Equation 9.1 is

$$E(x,y,z) = E(x,y,0)\exp(-i\phi) \tag{9.2}$$

where $\phi = k\int_0^z n(x,y,z')dz'$ Thus, propagation through this section of extended medium can be approximated by propagation through a phase screen that imposes the phase shifts $\phi(x,y)$. Note that this uses what is essentially a geometric optics approximation: the refractive index has been integrated along the path of a ray. A similar approximation can be used to describe spherical wave propagation, with the rays diverging rather than parallel.

The correlation function of the phase screen is given by a double integral over the three-dimensional correlation function of the refractive index.

$$\left\langle \phi_1\phi_2 \right\rangle = k^2 \int_0^z \int_0^z \left\langle n_1(z')n_2(z'') \right\rangle dz'dz'' \tag{9.3}$$

If n is a stationary random process the variables of integration can be converted to sum and difference coordinates, and the integral over the sum coordinate can be carried out to give

$$\left\langle \phi_1\phi_2 \right\rangle = 2k^2 \int_0^z \rho_n\left(\vec{r}_1 - \vec{r}_2, u\right)\left[z - u\right]du \tag{9.4}$$

where $\vec{r}$ is a vector in the plane of the phase screen, and ρ_n is the correlation function of the refractive index. Often the correlation length of the refractive index variations will be much less than z and the main contribution to the integral in Equation 9.4 will be from the region where the difference coordinate u is small. In this case the upper limit in the integral can be taken to be infinity, and the term $-u$ in the kernel neglected. This gives

$$\left\langle \phi_1\phi_2 \right\rangle = 2zk^2 \int_0^\infty \rho_n\left(\vec{r}_1 - \vec{r}_2, u\right)du \tag{9.5}$$

which results in a particularly simple form for the power spectral density of the phase screen in terms of the power spectral density of the refractive index variations

$$S_\phi\left(\kappa_x, \kappa_y\right) = zk^2 S_n\left(\kappa_x, \kappa_y, 0\right) \qquad (9.6)$$

This phase screen approximation models only the phase fluctuations induced by the varying refractive index. However, an approximation to the intensity fluctuations can be obtained by placing the screen at $z = 0$ and allowing the intensity fluctuations to develop by free-space propagation to z, as discussed in Chapter 5. Clearly, this can overestimate the strength of the intensity fluctuations because all of the phase variations are concentrated at the beginning of the propagation path; in reality, those at the end of the propagation path will have little effect on intensity fluctuations. A better approximation can be obtained by placing the phase screen at the halfway point, that is, $z/2$ [6].

If Taylor's hypothesis can be assumed (see Section 5.3), the correlation function of the temporal fluctuations in phase introduced by uniform motion of the scattering layer are given by Equation 9.5 with $\bar{r}_1 - \bar{r}_2$ replaced by $v\tau$, v being the speed of the transverse motion. Taking the Fourier transform gives the spectrum as

$$W_\phi(\omega) = \frac{1}{2\pi v} \int_{-\infty}^{\infty} S_\phi\left(\frac{\omega}{v}, k_y\right) dk_y \qquad (9.7)$$

Although this approach neglects the phase fluctuations induced by diffraction and interference, it has been found to give good agreement with experimental data in propagation of a laser beam through atmospheric turbulence. An example is shown in Figure 9.1 where Equation 9.7 predicts a $-8/3$ power-law slope in the spectrum (see Section 9.5) and this is seen in the experimental data over more than two decades of frequency.

9.3 Power-Law Models for the Refractive Index Spectrum

Refractive index fluctuations in extended media are usually multiscale and can often be described by a power-law structure function, as discussed in Chapter 7. Consider a three-dimensional refractive index structure function with exponent a.

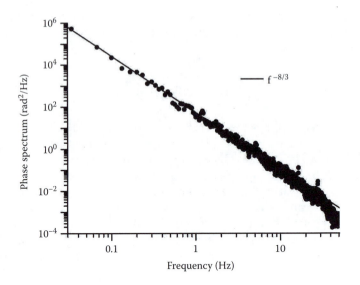

FIGURE 9.1
A measured phase spectrum from laser light propagated through atmospheric refractive index variations (previously unpublished data), solid line is −8/3 slope from Equation 9.17.

$$D_n(\bar{r}) = C_n^2 |\bar{r}|^a \tag{9.8}$$

The power spectral density is the Fourier transform of the autocorrelation function

$$S_n(\bar{\kappa}) = \iiint \rho_n(\bar{r}) \exp(-i\bar{r} \cdot \bar{\kappa}) dV \tag{9.9}$$

If the autocorrelation function depends only on the magnitude of $\bar{r}$, the variables of integration can be transformed to spherical polar coordinates r, θ, and ϕ, where θ is the angle between $\bar{r}$ and $\bar{\kappa}$. Using $dV = r^2\sin(\theta)drd\theta d\phi$, the θ and ϕ integrals can be evaluated to give

$$S_n(\bar{\kappa}) = \frac{4\pi}{|\bar{\kappa}|} \int_0^\infty r \sin(|\bar{\kappa}|r) \rho_n(r) dr \tag{9.10}$$

The autocorrelation function can be written in terms of the structure function as $\langle n^2 \rangle - D_n(r)/2$. The first term is undefined because it depends on the outer scale, which is not included in Equation 9.8. However, the second term gives rise to a power-law component in the power spectral density, which is

$$\frac{-2\pi C_n^2 \Gamma(a+2)}{|\bar{\kappa}|^{a+3}} \sin\left(\pi\left[1+\frac{a}{2}\right]\right) \tag{9.11}$$

A spectrum which has this same behavior but which includes an outer scale (i.e., a spatial outer scale, which gives a cutoff in the power law at low frequencies in the spectrum) is

$$S_n\left(\bar{\kappa}\right) = \frac{-2\pi C_n^2 \Gamma(a+2)\sin\left(\pi\left[1+\frac{a}{2}\right]\right)}{\left(|\bar{\kappa}|^2 + \left(\frac{2\pi}{l_o}\right)^2\right)^{\frac{a+3}{2}}} \tag{9.12}$$

The functional form of the outer scale in this equation, although commonly used, is arbitrary and chosen for mathematical convenience. The outer scale parameter l_o, discussed in Section 7.1, introduces a cutoff at large spatial scales and thus gives a finite variance for the refractive index fluctuations. If the condition applied in deriving Equation 9.5 (that the correlation length is less than the medium length) applies, then Equation 9.12 is also the functional form of the equivalent two-dimensional phase screen spectrum. From this the correlation function of the phase screen can be obtained by inverse Fourier transformation.

$$\rho\left(\bar{r}\right) = \frac{k^2 z}{4\pi^2} \int_{-\infty}^{\infty}\int_{-\infty}^{\infty} S_n\left(\bar{\kappa}\right) e^{-i\bar{\kappa}.\bar{r}} d^2\kappa \tag{9.13}$$

Since S_n is a function of the modulus of $\bar{\kappa}$ only, the integral can be converted to polar coordinates and the angular integral evaluated to give

$$\rho\left(r\right) = \frac{k^2 z}{2\pi} \int_0^{\infty} S_n\left(\kappa\right) J_0\left(\kappa r\right) \kappa d\kappa \tag{9.14}$$

where κ and r are the moduli of $\bar{\kappa}$ and $\bar{r}$, respectively. The structure function of the phase screen is given by

$$D\left(r\right) = 2\left(\rho\left(0\right)-\rho\left(r\right)\right) = \frac{k^2 z}{\pi} \int_0^{\infty} S_n\left(\kappa\right)\left[1 - J_0\left(\kappa r\right)\kappa d\kappa\right] \tag{9.15}$$

For separation r much less than l_o, the outer scale in Equation 9.12 does not affect the structure function, which becomes

$$D(r) = \frac{C_n^2 k^2 z \sqrt{\pi} \Gamma\left(-\dfrac{a+1}{2}\right)}{\Gamma\left(-\dfrac{a}{2}\right)} |r|^{a+1} \qquad (9.16)$$

In deriving this formula the reflection and duplication formulae for the gamma function have been used [7]. Thus a three-dimensional power-law structure function of exponent a is represented by a two-dimensional phase screen with a structure function of exponent $a+1$. The phase fluctuation spectrum can be found by substituting Equation 9.12 into Equation 9.7, which gives

$$W_\phi(\omega) = \frac{-\pi C_n^2 z k^2 2^{a+1} \Gamma\left(\dfrac{a+2}{2}\right)}{v \Gamma\left(-\dfrac{a}{2}\right)\left(\dfrac{\omega^2}{v^2} + \left(\dfrac{2\pi}{l_o}\right)^2\right)^{\frac{a}{2}+1}} \qquad (9.17)$$

It can be seen that the outer scale in the spatial structure function introduces a low frequency cutoff in the spectrum. This is because large spatial scales produce the slowest fluctuations and thus appear in the low frequency part of the spectrum. At higher frequencies, the spectrum has a power-law dependence on frequency, with exponent $-(a+2)$. For the turbulent atmosphere the power-law exponent is $a = 2/3$, which gives the $-8/3$ power law seen in Figure 9.1.

9.4 Multiple Phase Screens

An approximation which is better than the single phase screen can be obtained by dividing the propagation region into a number of short layers, each one of which is modelled by a phase screen followed by a free space path. If, as in the development of Equation 9.5 and Equation 9.6, the correlation length is less than the interscreen distance, the screens can be taken to be uncorrelated. This is the basis of numerical modelling by multiple phase screens, discussed in Section 14.8, and has also been used to investigate the variation of the fourth moment of the field with propagation distance [8].

The simplest analytical approach is to assume that the modification to the field introduced by each phase screen is independent of the other screens, that is, to neglect multiple scatter (a more detailed discussion of the approximations used in this section can be found in Reference [9]). To see how this works, we will consider the spectrum of the phase fluctuations. The results of Equation 9.7 and Equation 9.17 neglect all wave optics effects. Let the same region of refractive index variations be modelled by multiple screens, each one of which is weak. Consider a field E_0 incident on the first phase screen. In the absence of further screens, the field E_1 at z (the end of the region) is given by the Huyghens-Fresnel integral

$$E_1\left(\bar{r}_1, z\right) = \frac{ik}{2\pi z} \int E_0\left(\bar{r}_0\right) e^{i\phi\left(\bar{r}_0\right)} \exp\left(\frac{-ik\left|\bar{r}_1 - \bar{r}_0\right|^2}{2z}\right) d^2\bar{r}_0 \qquad (9.18)$$

If ϕ is sufficiently small over the region of the phase screen that contributes to the integral, the exponential term containing ϕ can be expanded and only the first two terms retained. This gives

$$E_1\left(\bar{r}_1, z\right) = E_1^u\left(\bar{r}_1, z\right) - \frac{k}{2\pi z} \int E_0\left(\bar{r}_0\right) \phi\left(\bar{r}_0\right) \exp\left(\frac{-ik\left|\bar{r}_1 - \bar{r}_0\right|^2}{2z}\right) d^2\bar{r}_0 \qquad (9.19)$$

The first term is the field that would be produced at z in the absence of the phase screen; that is, the unscattered part. If the second term is small compared to the first, the main contribution to phase fluctuations comes from the component which is in quadrature to the unscattered part. Taking the incident field to be real, this is the imaginary part, which is

$$\theta_1 = \frac{k}{2\pi z} \int E_0\left(\bar{r}_0\right) \phi\left(\bar{r}_0\right) \sin\left(\frac{k\left|\bar{r}_1 - \bar{r}_0\right|^2}{2z}\right) d^2\bar{r}_0 \qquad (9.20)$$

The neglect of multiple scatter means that only the unscattered part in Equation 9.19 is taken to interact with the second phase screen, and so on for all the phase screens. The final phase is thus

$$\theta_{tot} = \sum_j \theta_j \qquad (9.21)$$

where each θ_j is the phase change that would be produced at z by the *j*th phase screen on its own. The sum is analogous to the integral in Equation

9.2; successive phase changes are introduced by a multiplication to the field, which results in an addition to the exponent. In the phase spectrum, the spectral powers resulting from each screen are simply added because the screens are taken to be uncorrelated. To see how this affects the phase spectrum, consider, for simplicity, a one-dimensional case where the phase screen is corrugated and the incident field has a Gaussian profile in the x direction, but is of infinite extent in the y direction. The phase will be calculated at the point $(0,0,z)$, the detection point. The y integral can be carried out, giving

$$\theta_1 = \left(\frac{k}{4\pi z}\right)^{\frac{1}{2}} \int_{-\infty}^{\infty} \phi(x_0) e^{-\frac{x_0^2}{D^2}} \left(\sin\left(\frac{kx_0^2}{2z}\right) + \cos\left(\frac{kx_0^2}{2z}\right)\right) dx_0 \qquad (9.22)$$

where D is the width of the Gaussian beam. This is for the first screen in the sequence, but the results for the others are entirely analogous. If the phase screen is moving with speed v in the x direction, $\phi(x_0)$ can be replaced by $\phi(x_0 + v\tau)$ and a Fourier transform with respect to τ gives

$$\bar{\theta}_1(\omega) = \bar{\phi}\left(\frac{\omega}{v}\right)\left(\frac{k}{4\pi z v^2}\right)^{\frac{1}{2}} \int_{-\infty}^{\infty} \cos\left(\frac{\omega x_0}{v}\right) e^{-\frac{x_0^2}{D^2}} \left(\sin\left(\frac{kx_0^2}{2z}\right) + \cos\left(\frac{kx_0^2}{2z}\right)\right) dx_0$$

$$(9.23)$$

The bar in Equation 9.23 indicates the Fourier transform of the variable under it. Note that, strictly speaking, the Fourier transform of ϕ does not exist; it is being used here as an intermediate step in calculating the power spectral density, along the lines of Equation 1.23 and Equation 1.24. Evaluating the integral and taking the modulus squared leads to a result for the power spectral density of θ in terms of that of ϕ (note that in applying the limit of Equation 1.25 an extra multiplying factor of v is needed to relate the temporal transform to the spatial one).

$$S_{\theta_1}(\omega) = \frac{W_\phi(\omega)}{N} \frac{s}{2\sqrt{1+s^2}} \left(\cos(\Psi) - \sin(\Psi)\right)^2 \exp\left(-\frac{\omega^2 D^2}{2v^2(1+s^2)}\right) \qquad (9.24)$$

where $s = \dfrac{kD^2}{2z}$, $\Psi = \left(\dfrac{\omega D}{2v}\right)^2 \dfrac{s}{1+s^2} - \dfrac{1}{2}\tan^{-1}(s)$, and N is the number of phase screens representing the extended medium. Thus, the spectrum of Equation 9.7, which was based on the approximation of Equation 9.2, is modified by being multiplied by a frequency dependent term; this is equivalent to a filtering operation in the time domain. The significance of this is revealed by looking at some limiting cases of Equation 9.24. Taking $k \to \infty$, which means neglecting wave optics effect, recovers the result in Equation 9.7, the geometric optics limit. In this limit, the Gaussian profile of the beam has no influence on the phase spectrum because the phase fluctuations arise simply from optical path length variations along a single line in the z direction which is centered on the detection point. When wave optics effects are introduced, diffraction becomes significant; small inhomogeneities scatter incident light away from the straight line of geometric optics. It might be thought that this would lead simply to a reduction in the power of high frequency components; inhomogeneities on small spatial scales will scatter light away from the detection point and thus contribute less to the second term in Equation 9.19. However, the actual effect is more complicated than this, because diffraction also means that light travelling on parallel ray paths is scattered and contributes to the field at the detection point. This can be seen by taking the limit of $D \to \infty$, when Equation 9.24 reduces to the plane wave result

$$S_{\theta_1}(\omega) = \frac{W_\phi(\omega)}{N} \frac{1}{2}\left(1 + \cos\left(\frac{\omega^2 z}{v^2 k}\right)\right) \tag{9.25}$$

The result of combining the different scattered components is to produce oscillations in the spectrum at high frequencies as they go in and out of phase. For finite D, the Gaussian beam introduces a cutoff frequency because it limits the angle from which diffracted light can reach the detection point. This cutoff frequency comes in via the Gaussian term in Equation 9.24 and moves to lower frequencies as D is reduced. In the limit in which diffraction is the dominant effect, the cutoff angular frequency becomes equal to $v/2D$. This corresponds to the reciprocal of the time taken for a refractive index inhomogeneity to cross the beam (this is similar to the effect seen in the case of particle scattering, discussed in Section 4.7).

The result for the extended medium is found by substituting the various terms of form in Equation 9.24 in the summation (Equation 9.21). In plane wave limit this gives

$$S_{\theta_{tot}}(\omega) = \frac{W_\phi(\omega)}{2}\left(1 + \sum_{j=0}^{N-1} \cos\left(\frac{\omega^2 z}{v^2 k}\left(1 - \frac{j}{N}\right)\right)\right) \tag{9.26}$$

The summation tends to smooth out the oscillations because the different terms oscillate at different rates. If the summation is approximated by an integral over z, the result is a sinc function, which gives the curve plotted in Figure 9.2 for the multiplying factor. In the plane wave case the effect of diffraction in this weak scattering approximation is not dramatic; a reduction by a factor of 2 in going from low to high frequencies. Of more practical consequence is the high frequency cutoff introduced by a Gaussian beam profile. Results for the two-dimensional calculation with a Gaussian beam are given in [10].

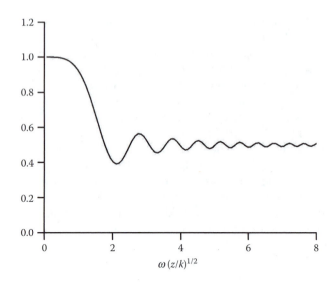

FIGURE 9.2
A first-order modification to the spectrum of a plane wave that has propagated through an extended medium; this function multiplies the geometric optics result; main effect of including diffraction is reduction by a factor of two at high frequencies.

The approximation made in Equation 9.20 assumes that the addition of fields results in the addition of their phases, which is clearly only valid when the phases themselves are small. Little is known about what happens to the phase spectrum in stronger scattering, but one would expect that the effects would initially appear at high frequencies because the large-scale variations in refractive index that produce the low frequencies in the phase spectrum will be least affected by wave optics. In very strong scattering one would expect the Gaussian limit for the field to be approached and the phase derivative to follow a student's t distribution as discussed in Section 2.8. (By contrast, the analysis leading to Equation 9.26 would predict a Gaussian distribution for the phase derivative.) The phase spectrum could then be obtained by dividing the Fourier transform of Equation 2.47 by ω^2 (the process of differentiation in the time domain being equivalent to multiplying by frequency in the frequency domain), and would depend on the correlation function of the field only.

9.5   Propagation of Electromagnetic Waves through Turbulence

This is the most widely investigated extended-medium problem. It is particularly important in the propagation of laser beams through the atmosphere. Refractive index variations in the atmosphere are caused by variations in temperature and, to a lesser extent, humidity. For optical wavelengths (i.e., visible and infrared), humidity variations are only significant near water surfaces [11] and are not considered here.

To a good approximation, the refractive index is related to temperature and pressure by [11]

$$n = 1 + 79 \times 10^{-6} \times \frac{P}{T} \qquad (9.27)$$

P is the pressure in millibars and T the temperature in degrees Kelvin. The important fluctuations are those in temperature.

During the day, solar heating injects energy into the atmosphere. The atmosphere is in a continual state of motion which mixes hot and cold air together to produce complex variations in local air temperature. Indeed, the motion of the atmosphere is often in a turbulent state and refractive index variations are sometimes simply referred to as "atmospheric turbulence."

Atmospheric turbulence has a number of detrimental effects on laser beam propagation. Beam spreading reduces the average intensity. Beam wander and scintillation both contribute to intensity fluctuations, which are a serious problem for laser communication systems. Phase fluctuations reduce the performance of coherent detection systems [12] and also degrade image quality [9]; the latter is also a problem in imaging systems that use ambient light, as is commonly observed when trying to see through "heat haze" on a hot day.

The usual approach to analyzing refractive index fluctuations is to assume that they conform to a simple model called Kolmogorov turbulence. This model says that for certain length scales the refractive index structure function follows a pure power law with a 2/3 exponent. The constant of proportionality that describes the strength of the fluctuations, C_n^2, is called the refractive index structure constant and has units of $m^{-2/3}$. The refractive index structure function is thus

$$D_n(\bar{r}) = C_n^2 |\bar{r}|^{2/3} \qquad (9.28)$$

The refractive index structure constant is related to a temperature structure constant, via Equation 9.27, by

$$C_n^2 = (79P / T^2 \times 10^{-6})^2 C_T^2 \qquad (9.29)$$

Maximum values of C_n^2 occur near to the ground and are on the order of $10^{-13} m^{-2/3}$. Minimum values occur in neutral stability conditions and may be as low as $10^{-16} m^{-2/3}$, even quite close to ground level [13]. In general, C_n^2 decreases quite rapidly with altitude.

The 2/3 power-law behavior of the refractive index structure function, which arises from simple dimensional arguments, is only expected for a certain range of $|\vec{r}|$ values, the so-called inertial regime. The upper and lower limits of this range are the outer and inner scales. The inner scale length is determined by the point at which viscous dissipation halts the energy cascade of the turbulent flow. This parameter has some variability and is on the order of millimetres near ground level [9]. For separations less than the inner scale, the structure function falls faster than the 2/3 law, and there is experimental evidence for a $|\vec{r}|^2$ dependence in this regime [14]. The outer-scale cutoff distance is determined by the size of the region of homogeneity. This is usually taken to be equal to the height above the ground at low altitudes, but is presumably governed by meteorological factors at higher altitudes. This theoretical model of the atmosphere has, therefore, three parameters, C_n^2 being the most important. All three can vary with time or location.

The 2/3 exponent in Equation 9.28 can be substituted into Equation 9.12 to give a refractive index power spectrum of [9]

$$S_n\left(\vec{\kappa}\right) = \frac{8.187 C_n^2}{\left(\left|\vec{\kappa}\right|^2 + \left(\dfrac{2\pi}{l_o}\right)^2\right)^{\frac{11}{6}}} \exp\left(-\left|\vec{\kappa}\right|^2 \left(\dfrac{l_i}{5.92}\right)^2\right) \tag{9.30}$$

known as the von Karman spectrum. Here, an inner scale has been introduced in the form of a Gaussian cutoff at high spatial frequencies. Note that if the factor of $(2\pi)^3$ is put in the transform, as opposed to the inverse transform, the multiplying factor in Equation 9.30 becomes 0.033, a form in which it is often encountered (in References [5] and [9], for example. Note that Equation 9.30 follows the definition of inner and outer scales in Reference [9]; Reference [5] uses different multiplying factors).

For the phase screen representation, the structure function can be found by putting $a = 2/3$ in Equation 9.16, giving

$$D\left(r\right) = 2.91 k^2 z C_n^2 r^{\frac{5}{3}} \tag{9.31}$$

The phase spectrum of Equation 9.17 has a region with a $-8/3$ power-law exponent, as seen in the experimental data of Figure 9.1.

9.6   Intensity Fluctuations

As a wave propagates through an extended medium, intensity fluctuations build up in a similar fashion to those produced by a deep phase screen; the second moment increases to a maximum and then decreases (as in Figure 6.4). At first sight it may seem surprising that the strength of fluctuations decreases at the same time as increasing phase fluctuations are being imposed by new parts of the extended medium. However, the situation is the same as with a phase screen illuminated by a plane wave; contributions from new regions of the screen appear with increasing distance and the addition of new components results in an approach to Gaussian statistics and a scintillation index of unity (Section 5.6). Experimental observations on laser beams propagating through atmospheric turbulence show that maximum scintillation-index values can be high; values in excess of four have been frequently reported [15–18], which seems at odds with power-law behavior, as pure power-law phase screens do not give such large values. This suggests that the inner spatial scale of the turbulence is important. Nevertheless, the simplest approach to analyzing intensity fluctuations in the region prior to the peak in the scintillation curve neglects the inner scale. The result for the plane wave scintillation index is [5]

$$S^2 = \exp(1.23 C_n^2 k^{7/6} L^{11/6}) - 1 \qquad (9.32)$$

This is calculated using the Rytov approximation, which is based on a series expansion in terms of the *logarithm* of the field [9]. The approximation works in a similar way to the phase spectrum calculation in Section 9.4, but is less easily justified where intensity fluctuations are concerned. This approach is believed to be valid for values of the scintillation index less than about 0.7, beyond which Equation 9.32 is an overestimate, an effect referred to as saturation.

The Rytov approximation leads naturally to the use of the lognormal distribution of Section 3.8 for the probability density of the intensity fluctuations. This arises as a result of the addition of many independent components in the logarithm of the field, leading to complex Gaussian statistics for this quantity. Thus, the intensity is given by

$$I = I_0 \exp\left[\sum_{n=0}^{N} x_n\right] \qquad (9.33)$$

If N is large in this expression, then the exponent will tend to become a Gaussian variable and the intensity can be represented in the form

$$I = I_1 \exp(X), \quad P(X) = \frac{1}{\sqrt{2\pi\sigma^2}} \exp\left[-\frac{X^2}{2\sigma^2}\right] \tag{9.34}$$

A factor of $\exp(\bar{X})$ has been absorbed in I_1, $\bar{X}$ being the average value of the summation and X the fluctuation about this value. The average of X is zero and its mean square value is $\sigma^2 \ll \bar{X}^2$. The scintillation index is given by $S^2 = \exp(\sigma^2) - 1$.

Figure 9.3 shows examples of the lognormal distribution with the intensity normalized by its mean value for scintillation indices of 0.04, 0.1, and 0.7 (the last being the limit of validity of the Rytov approximation), along with some experimental data. The weak scatter nature of this distribution can be seen in the fact that it does not allow for zeros in the intensity.

The lognormal distribution has also been found to apply to the case when light is gathered over a large area so that there is strong spatial integration of the fluctuations [19]. In this case, the applicability of the lognormal distribution is not limited to the weak scattering regime.

Many probability distributions have been proposed to describe intensity fluctuations in the saturation regime. It should be borne in mind that in the real world it is perhaps unrealistic to expect any single distribution to provide an exact description of the fluctuations under all conditions. It is rare to find turbulence that is uniform along the beam propagation path. Turbulence characteristics will vary in a complicated fashion under the influence of local variations in the topography and physical nature of the ground over which the beam is propagating; even for vertical paths, turbulence varies in a nonuniform manner, with many layers of enhanced fluctuations [13]. The diameter and divergence of the laser beam will also affect the distribution of the fluctuations, as will the position of the detector relative to the center of the beam.

In Reference [16] the K-distribution, discussed in Section 3.7 and Chapter 12, was found to fit experimental data in the saturation regime as the Gaussian limit is approached (see Figure 9.4). Note that the scintillation index must exceed unity, so it is not an appropriate distribution for weak turbulence. The K-distribution describes a compound process, and a number of other compound processes have been proposed as models for intensity scintillation. In Reference [17], a negative exponential variable whose mean is a lognormal variable was found to give a somewhat better fit to experimental data than the K-distribution. This log-normally modulated exponential variable has a density function given by an integral

$$P(I) = \frac{1}{\sqrt{2\pi\sigma_z}} \int_0^\infty \exp\left(-\frac{I}{z} - \frac{\left(\ln z + \sigma_z^2/2\right)^2}{2\sigma_z^2}\right) \frac{dz}{z^2} \tag{9.35}$$

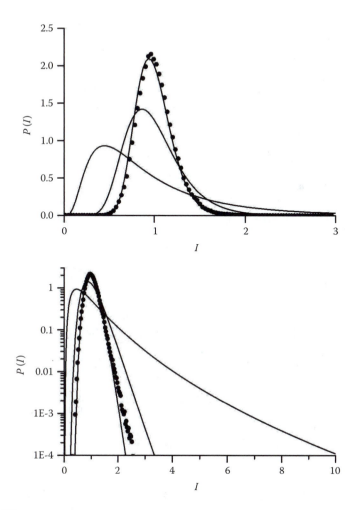

FIGURE 9.3
The lognormal distribution, unit mean, for S^2 of 0.04, 0.1, and 0.7; first curve is compared to some experimental data of equal scintillation index (using setup described in Reference [22]); lower graph replots data on logarithmic scale to better show behavior in wings of distribution.

The moments are

$$\langle I^n \rangle = n! \exp\left(\tfrac{1}{2} n (n-1) \sigma_z^2\right) \tag{9.36}$$

This is compared with the K-distribution in Figure 9.5, where both distributions have an S^2 value of two. It can be seen that there is some difference between the two at low intensity values, where the K-distribution probability density is higher. As I goes to zero, the log-normally modulated exponential probability goes to $(S^2+1)/2$, while the K-distribution goes to $2/(3-S^2)$ for $S^2 < 3$, and to infinity for values greater than or equal to three. At

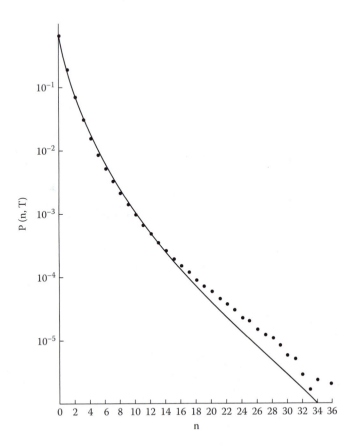

FIGURE 9.4
Comparison of experimental data with K-distribution; S^2 is 2.29 [23].

intermediate intensity values the distributions are in quite close agreement, but their asymptotic values differ, with the log-normally modulated exponential having higher values at large I.

A distribution which behaves like the lognormal for low scintillation and the log-normally modulated exponential for high scintillation, Beckmann's distribution, is discussed in this context in [20]. It is a compound distribution produced from a Rice variable, the mean of which is log-normally distributed. Beckmann's distribution has an extra parameter compared to the previously discussed distributions, which gives it the freedom to have two limiting forms. Two other distributions which have two parameters, in addition to the mean intensity, and which have been used to describe intensity fluctuations, are the I-K distribution discussed in Chapter 12 and the distribution of Equation 3.24 which is used extensively in Reference [4] and referred to there as the Gamma-Gamma pdf. The I-K distribution reduces to the K-distribution in strong scintillation, and can also describe weak scintillation, but its functional form differs from the lognormal in this regime

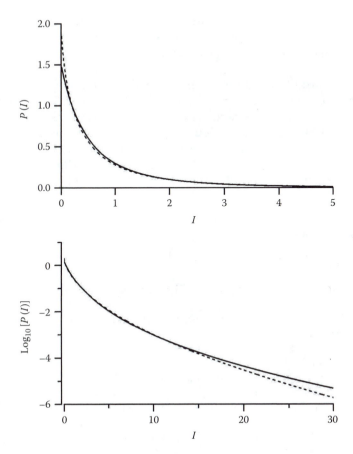

FIGURE 9.5
Comparison of log-normally modulated exponential distribution (solid line) with *K*-distribution (dashed line) for S² of two.

[21]; it also has the disadvantage of an unphysical cusp in the density function which occurs at the transition between the two functional forms in Equation 12.22.

Clearly the Gamma-Gamma includes the *K*-distribution of Equation 3.25 as a special case, as well as the gamma distribution of Equation 3.13. Its moments are

$$\frac{\langle I^n \rangle}{\langle I \rangle^n} = \frac{\Gamma(n+\alpha)\Gamma(n+\beta)}{(\alpha\beta)^n \, \Gamma(\alpha)\Gamma(\beta)} \tag{9.37}$$

It is also applicable to a *K*-distributed intensity which is spatially or temporally averaged, that is, by the collection of radiation over a finite aperture or by filtering of a detector output. This is because averaging of the small-scale speckle-like fluctuations gives a gamma process which is modulated by the gamma distributed large-scale fluctuations (see Chapter 12 and Section 13.4)

Of course, just fitting pdfs to experimental data points does not necessarily give any more information than that contained in the data itself. In order to predict intensity pdfs for situations that have not been measured, one needs to be able to calculate the distribution parameters from more fundamental quantities — models along the lines of Equation 9.30, for example. This has been done for the Gamma-Gamma pdf in Reference [4] although, at the time of writing, the resulting formulae have not been verified by detailed experimental measurements.

References

1. Tatarskii, V.I. *Wave Propagation in a Turbulent Medium.* New York: McGraw-Hill, 1961.
2. *Laser Beam Propagation in the Atmosphere.* J.W. Strohbehn, ed. Heidelberg: Springer-Verlag,1978.
3. Ishimaru, A. *Wave Propagation and Scattering in Random Media.* New York: Academic Press, 1978.
4. Andrews, L.C., R.L. Phillips, and C.Y. Hopen. *Laser Beam Scintillation with Applications.* SPIE, 2001.
5. Fante, R.L. Electromagnetic Beam Propagation in Turbulent Media. *Proc. IEEE* 63 (1975): 1669–92.
6. Booker, H.G., J.A. Ferguson, and H.O. Vats. Comparison between the Extended Medium and the Phase Screen Scintillation Theories. *J. Atmos. Terrest. Physics* 47 (1985): 381–99.
7. Abramowitz, M., and I.A. Stegun. *Handbook of Mathematical Functions.* New York: Dover, 1971.
8. Uscinski, B.J. Analytical Solution of the Fourth-Moment Equation and Interpretation as a Set of Phase Screens. *J. Opt. Soc. Am.* A 12 (1985): 2077–91.
9. Goodman, J.W. *Statistical Optics.* New York: Wiley, 1985.
10. Ishimaru, A. Chapter 5 in Reference [2].
11. Beland, R.R. Chapter 2 in IR/EO Handbook Volume 2, *Atmospheric propagation of radiation.* SPIE, 1993.
12. Frehlich, R.G., and M.J. Kavaya. Coherent Laser Radar Performance for General Atmospheric Refractive Turbulence. *Appl. Opt.,* 30 (1991): 5325–52.
13. Good, R.E., R.R. Beland, E.A. Murphy, J.H. Brown, and E.M. Dewan. Atmospheric Models of Optical Turbulence, in Modeling of the Atmosphere. *Proc SPIE 928 (1988):* 165–86.
14. Lukin, V.P., and V.V. Pokasov. Optical Wave Phase Fluctuations. *Applied Optics* 20 (1981): 121–35.
15. Phillips, R.L., and L.C. Andrews. Measured Statistics of Laser-Light Scattering in Atmospheric Turbulence., *J. Opt. Soc. Am.* 71 (1981): 1440–45.
16. Parry, G., and P.N. Pusey. K-Distributions in Atmospheric Propagation of Laser Light. *J. Opt. Soc. Am.* 69 (1979): 796–98.

17. Churnside, J.H., and R.J. Hill., Probability Density of Irradiance Scintillations for Strong Path-Integrated Refractive Turbulence. *J. Opt. Soc. Am.* A4 (1987): 727–33.
18. Consortini, A., F. Cochetti, J.H. Churnside, and R.J. Hill. Inner Scale Effect on Intensity Variance Measured for Weak to Strong Atmospheric Scintillation. *J. Opt. Soc. Am.* A17 (1993): 2354–62.
19. Hill, R.J., J.H. Churnside, and D.H. Sliney. Measured Statistics of Laser Beam Scintillation in Strong Refractive Turbulence Relevant to Eye Safety. *Health Physics* 53 (1987): 639–47.
20. Hill, R.J., and R.G. Frehlich. Probability Distribution of Irradiance for the Onset of Strong Scintillation. *J. Opt. Soc. Am.* A14 (1997): 1530–40.
21. Churnside, J.H., and R.G. Frehlich. Experimental Evaluation of Log-Normally Modulated Rician and Ik Models of Optical Scintillation in the Atmosphere. *J. Opt. Soc. Am.* A6 (1989): 1760–66.
22. Ridley, K.D., E. Jakeman, D. Bryce, and S. Watson. Dual-Channel Heterodyne Measurements of Atmospheric Phase Fluctuations. *Appl. Optics* 42 (2003): 4261–68.
23. Parry, G. Measurement of Atmospheric Turbulence Induced Intensity Fluctuations in a Laser Beam. *Optica Acta* 28 (1981): 715–28.

10

Multiple Scattering: Fluctuations in Double Passage and Multipath Scattering Geometries

---

10.1 Introduction

Apart from the extended medium propagation problem considered in the previous chapter, the scattering models that have been studied so far have only involved *single* scattering by particles or phase variations. In the previous chapter, the weak scattering treatment was also effectively a single scattering approximation, but in the saturated regime *multiple* scattering by a sequence of inhomogeneities had to be taken into account [1]. Multiple scattering is a feature of light propagation through dense particle suspensions such as cloud and fog, encoding properties of the medium while at the same time limiting the transfer of information through it [2,3]. It is also a problem that has received a great deal of attention from the solid-state physics community interested in electron scattering in disordered media [4]. The subject has been studied for many years, but owing to the intractable nature of the calculations, attention has focused mainly on evaluating the most basic characteristic of the scattered (or transmitted) radiation, namely, its mean intensity. With the development of powerful computers, however, there has been renewed interest in multiple scattering phenomena and a large number of new investigations have been carried out (for a list of references see [5] for example).

Effects that are closely related to multiple scattering phenomena also arise in *double passage* scattering geometries that occur widely in remote sensing where the transmitted wave passes twice through the scattering medium: once on its outward path to the object of interest and once on its return to the detector [6–8]. Similar effects are also generated in *multipath* configurations where the radiation passes from the source to the scatterer and back to the observer by more than one route. This may occur, for example, when the scatterer is unavoidably illuminated via a nearby reflecting surface as well as directly [8].

The main concern of this volume is with modelling *fluctuations* in the scattered wave field, and there are indeed few analytical results that are useful in this context for situations where multiple scattering is present [5]. However, many double passage configurations can be described with the help of the phase screen models discussed in earlier chapters and may also be used to illustrate some of the important features of multiple scattering including the characteristic signature known as *enhanced backscattering*. Although analytical results for the statistical properties of the radiation scattered by these models are confined to a few limiting cases, they may be used to good effect in the performance analysis of practical remote sensing systems.

10.2 Multiple Scattering in Particulates

The earliest results on fluctuations of multiply scattered light appeared in the photon counting literature in the early 1970s and relate to the sequential *single* scattering of light through two particle suspensions [9]. If many particles contribute to the field scattered by the first system, then, in the far field, the second suspension is illuminated by a Gaussian speckle pattern. The key consideration then becomes the spatial and temporal coherence properties of this light. If the light is coherent over the volume of the second scattering system, that is, if the transverse speckle size and longitudinal coherence length $c\tau_c$ (τ_c being the fluctuation time of the light governed by the phase fluctuations in the source and the particle motion) define a volume that is larger than the size of the second sample, then this suspension is illuminated by a coherent field whose intensity is fluctuating according to a negative exponential distribution (Equation 2.20). The second scattering then imposes a second Gaussian modulation on the field variable, and the scattered intensity is a negative exponential distribution whose mean is modulated by another statistically independent negative exponential distribution. Since the negative exponential distribution is a member of the gamma distribution class (Equation 3.13), the distribution of the intensity of radiation that has been doubly scattered in this situation is a member of the K-distribution class (see, for example, Equation 3.25 and Equation 4.28)

$$P(I) = 2bK_0\left(2\sqrt{bI}\right) \tag{10.1}$$

This distribution indicates enhanced fluctuations, having a second normalized moment of four and higher moments given by

$$\frac{\langle I^n \rangle}{\langle I \rangle^n} = \left(n! \right)^2 \tag{10.2}$$

If, on the other hand, the speckle pattern illuminating the second suspension is *not* coherent over the scattering volume (for example, if the speckle size is much smaller than the area of the second suspension that contributes to the field at the detector), then each of the secondary scatterers merely introduces a random phase factor in addition to the one already present and the outcome remains a complex Gaussian process with a negative exponential distribution of intensity. When the radiation illuminating the second system is *partially* coherent, the negative exponential modulation introduced by the second scattering may be replaced approximately by a gamma distribution (Equation 3.13) with larger fluctuation parameter α [10]. This will again result in a *K*-distribution of the type (Equation 3.25) but with a larger index ($\alpha > 1$) so that the fluctuation distribution will lie somewhere between (Equation 10.1) and a negative exponential. Both experimental evidence and more detailed theoretical work suggest that these arguments can be applied more generally to multiple scattering suspensions in forward scattering configurations [5].

As remarked earlier, an additional effect characteristic of multiple scattering occurs in backscattering configurations. *Coherent* enhanced backscattering can be understood from Figure 10.1 that shows radiation paths through a simple particle scattering system. It is clear that multiple scattering paths that return in the direction of the source are reversible and will add constructively, while those that return in other directions will add with random phases. Therefore, any multiple scattering will always give rise to a bright region in the backscattering direction. Moreover, since this direction is independent of wavelength, the effect will be observable in broadband illumination (e.g., white light) when the usual single scattering speckle pattern is washed out. The size of the bright region is determined by the lateral extent of the reversible paths and thus by the particle density. Several naturally occurring phenomena, including the "glory" [11] and the "opposition effect" that is, the exceptional brightness of the full moon [12], have been attributed to enhanced backscattering as well as many surface scattering effects observed as long ago as the 1930s [13]. A similar effect in the scattering of electrons by disordered media known as "weak Anderson localization" is well known to the solid-state physics community [14].

The spatial and temporal spectra of multiply scattered waves are generally broader than those of singly scattered radiation. This feature is exploited as a new source of information in an experimental technique known as "diffusing wave spectroscopy." However it can also be used, through low-pass filtering, to discriminate against multiply scattered radiation to improve the interpretation of data. Many papers and a wealth of references are to be found in [15].

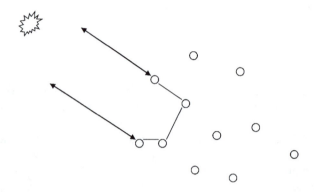

FIGURE 10.1
Enhanced backscattering in a particulate; optical paths in backscattering direction are reversible and add coherently.

10.3 Enhanced Backscattering through Continuous Media

In the case of scattering by continuous media, *incoherent* enhancement of the intensity may occur near the backscattering direction due to geometrical optics effects. The simplest man-made systems that generate such effects are corner reflectors and "cat's eyes" and the effects are exploited in reflective paints. However, it is not difficult to demonstrate experimentally that random scatterers may also produce incoherent enhanced backscattering. For example, when a mirror is illuminated through a piece of roughened glass both coherent and incoherent backscattering effects can be generated depending on the separation distance and the roughness characteristics of the glass. This can be understood with reference to Figure 10.2 and Figure 10.4.

The coherent effect illustrated in Figure 10.2 is generated in the same way as in the particle scattering system shown in Figure 10.1. The incident radiation is forward scattered by inhomogeneities in the glass, reflected at the mirror and is rescattered on the return journey. The mirror is sufficiently far behind the glass that an individual optical path passing through one scattering inhomogeneity on the outward trajectory returns through a different one. In the backscattering direction, two such paths may be traced in opposite directions from the source having equal length and these will add constructively as previously described. The size of the bright region around the backscattering direction is determined by two variables: the angular width of the scattering at the first pass and the mirror-scatterer separation. These variables govern the area of the ground glass that encompasses reversible paths, and this in turn determines the width of the enhancement peak according to the laws of diffraction. Figure 10.3 shows the appearance of the speckle pattern in an actual experiment with finely ground glass [16].

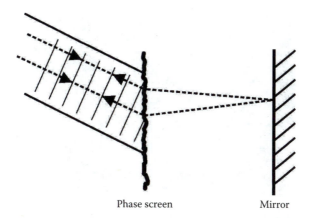

FIGURE 10.2
Coherent enhanced backscattering from a mirror illuminated through a phase screen.

When the mirror is sufficiently close to the roughened glass, as shown in Figure 10.4, some individual lens-like inhomogeneities may focus the scattered light onto the mirror plane. The effect is then similar to a cat's eye or corner reflector as all the rays through these random lenses are returned in the direction of incidence. This again gives rise to a peak in the backscattering direction that now reflects the *geometrical* properties of the scatterer. The width of this incoherent backscattering peak is determined by the average tilt of surface elements of the glass rather than by diffraction, and is therefore comparable to the angular distribution of scattered intensity. The speckle pattern obtained using obscuring bathroom glass is shown in Figure 10.5. Clearly the degree to which the roughness of the glass is "smoothly varying" will influence the magnitude of this incoherent enhancement.

10.4 A Phase Screen Model for Enhanced Backscattering

Double passage geometries of the kind described above, in which radiation passes through a random medium, strikes an object, and is returned to a detector in the source plane, are important because they are commonly encountered in radar, sonar, and lidar remote sensing systems. The example of light scattering from a mirror placed behind ground glass is a generic configuration that can be studied in more detail by adopting a phase screen model for the scatterer. Because of the double passage through the screen, a calculation of the mean intensity closely follows the second intensity moment calculations described in Chapters 5–8 and will not be given here in full. Using a similar combination of steepest descents and function modelling it may be shown that, [17] for the case of a deep, smoothly varying corrugated

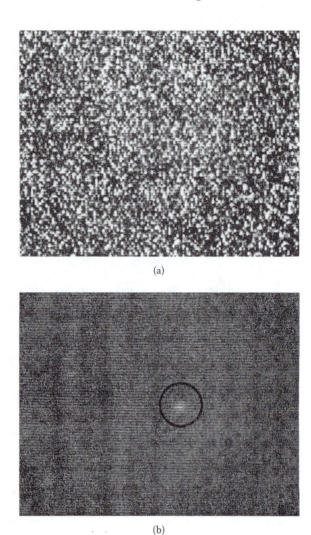

(a)

(b)

FIGURE 10.3
Speckle pattern observed when laser light is back scattered by mirror behind a ground glass phase screen, (a) stationary screen, (b) rotating screen showing coherent enhancement peak [16].

Gaussian screen with illuminated area much greater than the largest phase inhomogeneity, and with the mirror in the Fresnel region behind the screen

$$\left\langle I(\theta_{s}) \right\rangle = \frac{A}{m_0\sqrt{2}} I_1(Q)\exp\left(-P^2/2m_0^2\right) + \frac{A}{m_0} I_2(Q)\exp\left(-P^2/m_0^2\right) \quad (10.3)$$

In this formula

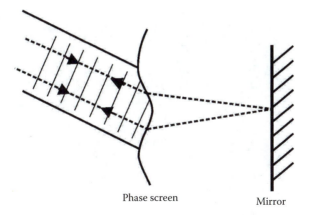

FIGURE 10.4
Geometrical mechanism for enhancement of intensity backscattered from a mirror illuminated through a phase screen (random cat's eye effect).

$$I_1(Q) = \frac{2q}{\sqrt{\pi}} \exp(-q^2) \left[Ci(2k\xi Q/\sqrt{3}) + \frac{1}{2} \sum_{n=0}^{\infty} \frac{\left(4q^2\right)^n}{2n!} \Gamma(n, 2q^2/m_0^2) \right]$$

$$I_2(Q) = 1 - \frac{1}{2}\left[erfc(q - Q\sqrt{2}/m_0) + erfc(q + Q\sqrt{2}/m_0) \right] + \hspace{1.5cm} (10.4)$$

$$+ \frac{l}{l_0} \exp\left(-2k^2 l^2 Q^2\right) Re\{erfc\left(ql_0/l - iklQ\sqrt{2}\right)\}$$

$Ci(x)$ is the cosine integral, $\Gamma(a, x)$ is the incomplete gamma function, and $erfc(x)$ is the complementary error function [18]; $q = \xi^2/2dh_0\sqrt{6}$, d is the mirror-screen separation, h_0 is the rms surface height, m_0 its rms slope, and ξ its characteristic correlation length (Equation 6.1 with $\varphi_0 = kh_0$). The angle dependence of the scattering governed by the result in Equation 10.4 is defined in terms of the factors

$$P = \frac{1}{2}\left(\sin\theta_i + \sin\theta_s\right), \quad Q = \frac{1}{2}\left(\sin\theta_s - \sin\theta_i\right) \hspace{1.5cm} (10.5)$$

and the scattering lengths are given by

$$l_0 = dm_0, \quad l^{-2} = W_0^{-2} + l_0^{-2} \hspace{2cm} (10.6)$$

where W_0 is the waist size of an illuminating Gaussian beam profile focused on the screen (see Equation 4.3). A result for the isotropic case is also to be found in Reference [17].

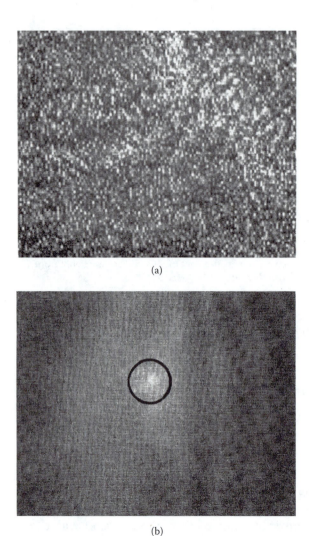

(a)

(b)

FIGURE 10.5
Speckle pattern observed when laser light is back scattered by a mirror in the focusing region behind a bathroom glass screen, (a) stationary screen, (b) rotating screen showing geometrical enhancement peak [16].

The two contributions in Equation 10.4 have specific physical significance. The first takes account of the lens-like behavior of individual inhomogeneities in the region between the screen and the mirror, and is rather similar to the single scatterer or non-Gaussian contribution to the second moment of intensity fluctuations in a single pass scattering geometry. The second gives the coherent contributions arising from paths that traverse different scattering elements, and is reminiscent of the interference or Gaussian contribution to the second moment of intensity fluctuations in a single pass scattering geometry. The limit $q \to \infty$ corresponds to a situation in which the two scattering events are superimposed, that is, to scattering by coincident identical screens. In this case, $I_2 \to 0$ and

$$\langle I \rangle \to \frac{A}{m_0 \sqrt{2}} \exp\left(-P^2/2m_0^2\right) \tag{10.7}$$

This is the result expected for a double strength single screen of mean square slope $2m_0^2$ (compare with result of Equation 6.26 by setting $\theta_i = 0$, $\theta_s = \theta$). In the opposite limit, $q \to 0$, the mirror is so far from the screen that the scattered field illuminating the screen on its return passage is a zero mean complex Gaussian process or fully developed speckle pattern. In this case $I_1 \to 0$ and

$$\langle I \rangle \to \frac{A}{m_0} \exp\left(-P^2/m_0^2\right)\left[1 + \frac{l}{l_0} \exp\left(-2k^2 l^2 Q^2\right)\right] \tag{10.8}$$

The first term in this expression is just the scattered intensity expected from a single screen. The second factor is a coherent enhancement due to constructive interference of radiation that has traversed reversible paths through the scattering configuration. The significance of the scattering lengths (Equation 10.6) now becomes clear: l_0 is the spread of radiation at the second pass and measures the maximum extent of the region traversed by reversible paths. When the illuminated area is much greater than this, $l = l_0$ and the full enhancement factor of two is realized in the backscattering direction, $Q = 0$. Away from this direction the enhancement falls off with an angular (diffraction) width $\propto (kl_0)^{-1}$ that extends over many speckles. However, when the illuminated area is smaller than l_0, the region of reversible paths is restricted, the enhancement is reduced, and the width of the enhancement peak decreases towards the speckle value $(kW_0)^{-1}$.

Figure 10.6 and Figure 10.7 show plots of the result of Equation 10.3 compared to data taken using ground glass and obscure bathroom glass screens respectively. The size of the incoherent backscattering peak increases with $\ln \varphi_0$ reflecting the divergence with wave number found in the second moment calculations of Chapter 6. Its width is governed by the geometrical spread of rays and is therefore comparable to the specular distribution of scattered intensity. Note that when the mirror lies near the focusing plane, l_0 is comparable to the size of the phase inhomogeneities and therefore by assumption $l_0 \ll W_0$. The ratio of the angular widths of the incoherent-to-coherent contributions to the backscatter enhancement is thus $m_0 (kl_0)^{-1} = km_0^2 d \approx \varphi_0$ in the focusing plane. Since by assumption $\varphi_0 \gg 1$, the residual coherent enhancement in this regime remains confined to a fairly narrow spike in the backward direction. Calculations of the spatial coherence function of the scattered field show that its decay continues to be determined by the usual aperture diffraction scale. Thus both incoherent and residual

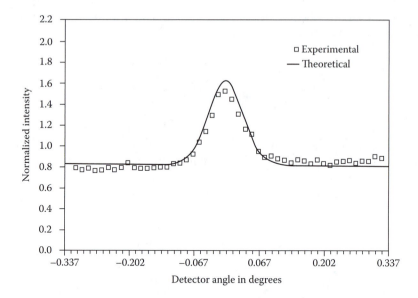

FIGURE 10.6
Coherent enhancement regime, comparison of data with theory (Equation 10.8) for a mirror-screen distance of 1 mm; $\theta_i = -3^0$ [16].

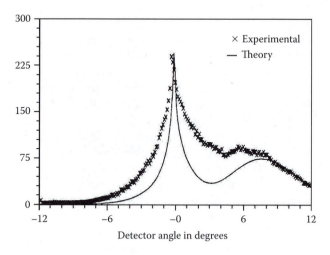

FIGURE 10.7
Incoherent enhancement regime, comparison of data with theory (Equation 10.3) for a mirror-screen distance of 2 mm; $\theta_i = -3^0$ [16].

coherent backscatter enhancement continue to extend over many speckles when the mirror is in the focusing plane of the screen.

The theoretical results explained above have also been tested experimentally in laboratory experiments in which laser light was passed through a turbulent plume of air above a gas burner and reasonable agreement obtained [19]. Outdoor observations of enhanced backscattering have been less widely reported but measurements made of both visible and infrared light that had propagated over 1 km folded paths through the turbulent atmosphere has confirmed that it is indeed a robust feature of such double passage geometries [20].

10.5 Fluctuations in Double Passage Configurations

The higher-order statistics of radiation scattered by the phase screen/plane mirror configuration remain Gaussian if the area of the screen contributing at the detector is greater than the speckle size at the screen following the first pass. Non-Gaussian statistics might be expected if (1) the area of the screen contributing at the detector is *less* than the speckle size at the screen following the first pass (when the considerations of Section 10.2 apply), or (2) when few inhomogeneities are illuminated. It should be emphasized that the first possibility is excluded by the restriction $kW_0^2/4d \gg 1$ assumed in the derivation of the result in Equation 10.3. The statistics in the second case are sensitive to the detail of the screen model. However, as previously remarked, the double passage geometry is common in remote sensing systems where the object of interest does not usually have the characteristics of a plane mirror. A more frequently encountered case open to simple analysis is shown in Figure 10.8. Here a point source illuminates a point reflector through an arbitrary extended random medium and observation of the returned signal takes place back in the source plane. By assumption, the forward scattered field is some scattering function $f(t)$ times the incident field

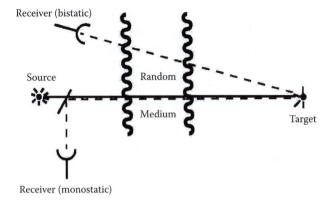

FIGURE 10.8
Double passage scattering geometry.

and is coherent over the reflector. If the detector is near the backscattering direction, then the returning radiation passes back through the same inhomogeneities and, invoking the principle of reciprocity, the received field is proportional to f^2 and the intensity is proportional to $|f|^4$. The distribution of intensity fluctuations at the source plane can therefore be calculated without difficulty, given that at the reflector. In particular, if the point reflector is illuminated by a Gaussian speckle pattern from the first pass, then the fluctuation distribution of the backscattered intensity is

$$P(I) = 2b\sqrt{\frac{2}{\pi}}\left(2bI\right)^{-\frac{1}{4}} K_{1/2}\left(2\sqrt{bI}\right) = \frac{\exp\left(-2\sqrt{bI}\right)}{\sqrt{2bI}} \qquad (10.9)$$

This distribution is another member of the K-distribution class (Equation 3.25) with normalized moments

$$\frac{\left\langle I^n \right\rangle}{\left\langle I \right\rangle^n} = \frac{(2n)!}{2^n} \qquad (10.10)$$

and, with a second normalized moment of six, it is also a member of the Weibull class of distributions of Equation 3.33 that have often been used to characterize non-Gaussian data, particularly microwave sea echo [21].

Now, if the detector is situated well away from the backscattering direction, then the radiation will pass through different inhomogeneities on its return path through the scattering medium so that $I \propto |f_1^2 f_2^2|$ where f_1 and f_2 are statistically independent realizations of the same scattering factor. When the reflector is illuminated by a Gaussian speckle pattern in this case, the returned intensity is governed by the K_0 distribution (Equation 10.1) as in the particle scattering configuration considered in Section 10.2.

If the point scatterer is replaced by a reflecting element of finite size in the configurations described above, then this will scatter radiation with a finite lateral coherence length. The properties of the received signal will then depend on whether radiation from more than one speckle in the reflection plane reaches the detector. If this is so and radiation on first passage is a complex Gaussian process, then the additional fluctuations imposed at the second pass will be reduced. Again, it is possible to use a gamma distribution with index greater than unity to model the effect on the returned intensity (as in the particle scattering case considered earlier), and the received signal will be K-distributed with an index that is increased by comparison with the fully coherent case (see Chapter 13).

The temporal and spatial fluctuation spectra of waves that have been scattered in double passage configurations depend on characteristics of the source and reflector. For example, in the case of ground glass illuminated by a beam of radiation in front of a plane mirror in the Fresnel region, $kW_0^2/4d \gg 1$, the characteristic scale of the scattered speckle pattern is unchanged by the double pass [17]. On the other hand, in the case of a point source/point reflector configuration the fluctuation spectra are generally broadened. In the geometry of Figure 10.8 for example, assuming that the radiation incident on the reflector is Gaussian, the field correlation function in the backscattering direction is given by

$$\left\langle EE'^* \right\rangle \propto \left\langle f^2 f'^{*2} \right\rangle = 2 \left\langle \left| f^2 \right| \right\rangle^2 \left[g^{(1)} \right]^2 \tag{10.11}$$

rather than $g^{(1)}$ expected for a single pass. On the other hand, in off-axis configurations

$$\left\langle EE'^* \right\rangle \propto \left\langle f_1 f_2 f_1'^* f_2'^* \right\rangle = \left\langle \left| f^2 \right| \right\rangle^2 \left[g^{(1)} \right]^2 \tag{10.12}$$

For the case of a Gaussian coherence function, $g^{(1)}$, these results would imply a characteristic fluctuation time/length that is reduced by a factor of $\sqrt{2}$ and a spectrum that is broadened by the same factor.

As indicated earlier in the chapter, multiple scattering is a source of new information. A tutorial illustration is provided by a light scattering phase screen experiment that demonstrated how different orders of multiple scattering may be generated and coded by the frequency spectrum of the scattered radiation [22]. In this experiment the phase screen is a partially reflecting ground glass disc set at a small angle to the back reflecting mirror: a slight modification of the configuration shown in Figure 10.2. Rotation of the screen causes the scattered light to fluctuate. However, incident light is now multiply reflected between the phase object and the mirror, acquiring an additional Doppler shift at each bounce. The emerging light therefore contains a set of frequencies relating directly to the number of scatters and these can be measured by heterodyne detection (Figure 10.9).

Double passage effects must be taken into account when analyzing the performance of remote sensing systems. Since double passage through a random medium would enhance the return signal in a monostatic configuration by comparison with a bistatic one, it might naively be assumed that the former type of system would perform better in practice. However, as demonstrated above, the *fluctuations* in the return may also be enhanced in the backscattering direction so that, although the average signal to noise ratio may be improved, the degree of fading is increased. The interplay of these factors in target detection is analyzed in Chapter 13 [23].

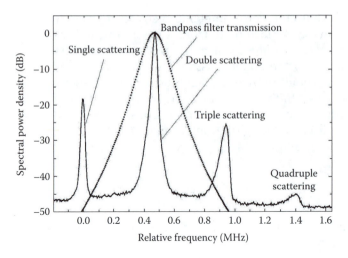

FIGURE 10.9
Power spectrum of light scattered through a rotating ground glass disc inclined at a small angle in front of a plane mirror [22].

10.6 Multipath Near Surfaces

The double passage scattering configuration illustrated in Figure 10.2 is one example of *multipath*. Another kind of multipath that is commonly encountered in remote sensing systems is illustrated in Figure 10.10. Here the object of interest lies close to a surface and may be illuminated directly or by reflection from the surface. Conversely, radiation may be scattered directly back to the detector or indirectly via the surface. The returned signal is thus

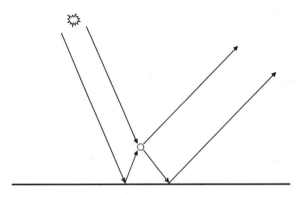

FIGURE 10.10
Multiple paths in scattering from an object near to a surface.

composed of four contributions that have traced different kinds of path. One will have encountered only the object of interest, one will have bounced on the surface before illuminating the object, one will have bounced on the surface after scattering by the object, and one will have bounced on the surface both before and after striking the object. The resulting field for a small spherical particle (or Huyghens point source) many wavelengths above a planar interface is given in terms of the reflection coefficient of the surface by

$$E \propto \left\{ 1 + R(\theta_i) \exp\left[i f(\theta_i) \right] \right\} \left\{ 1 + R(\theta_s) \exp\left[i f(\theta_s) \right] \right\} \qquad (10.13)$$

In this expression, R is the surface reflection coefficient corresponding to the incident and reflection angles, and $f(\theta) = 2kd \cos \theta$ is the phase shift relating to the height, d, of the scatterer above the surface, and measures the difference in length of the paths. It is clear that this expression predicts a set of fringes running parallel to the surface. This configuration is particularly relevant to the optical detection of contaminants on a surface, and can be exploited as a diagnostic tool by the semiconductor industry [24]. *Fluctuations* in the scattered intensity can occur due to a variety of effects. These include variations in particle size, distribution and height above the surface, and by variations in the character of the surface itself.

The model may easily be generalized to other remote sensing applications, such as radar detection near land or sea surfaces, by introducing an element of roughness on the surface and by integrating results for a point scatterer over the surface of the extended object of interest. In this case the reflection coefficients in Equation 10.13 will be stochastic scattering factors determined by the nature of the rough surface and the scattering geometry. For example, if the surface is rough compared to the wavelength and if many inhomogeneities contribute to the surface reflection, then R will be a complex Gaussian variable and each factor in the curly brackets in Equation 10.13 will be a Rice variable comprising the coherent sum of a constant component returned from the target particle and a Gaussian speckle pattern. In the backscattering direction the intensity is just the square of this kind of variable, and the intensity fluctuation distribution can be obtained directly from Equation 3.5 by making the appropriate transformation of variables. In other directions the scattered field will be the product of two independent Rice variables. In practice, scattering from rough surfaces often generates a non-Gaussian intensity pattern. However, the multipath intensity fluctuation distribution can be constructed in the same way for any given surface scattering model. Evidently, the considerations in these geometries are the same as that in the double passage geometries described in previous sections except for the addition of the coherent direct reflection from the target.

In the case of electromagnetic radiation, fluctuations also arise due to variations in particle shape and orientation, and the presence of the interface generally has a profound effect on the polarization characteristics of the signal that are important in many practical applications. A fuller description of the model shown in Figure 10.10 and its predictions in this case is given in the next chapter.

References

1. Ishimaru, A. *Wave Propagation and Scattering in Random Media*. New York: Academic Press, 1978.
2. Kuga, Y., and A. Ishimaru. Retroreflectance from a Dense Distribution Of Spherical Particles. *J. Opt. Soc. Am.* A1 (1984): 831–39.
3. Zege, E.P., A.P. Ivanov, and I.L. Katsev. *Image Transfer through a Scattering Medium*. Berlin: Springer-Verlag, 1991.
4. *Mesoscopic Phenomena in Solids*. Eds. B.L. Altshuler, P.A. Lee, and R.A. Webb. Amsterdam: North-Holland, 1991.
5. Garcia-Martin, A., J.J. Saenz, and M. Nieto-Vesperinas. Spatial Field Distributions in the Transition from Ballistic to Diffusive Transport in Randomly Corrugated Waveguides. *Phys. Rev. Letts.* 84 (2000): 3578–81.
6. Banakh, V.A., and V.L. Mironov. Lidar In A Turbulent Atmosphere. New York: Artech House, 1988.
7. Kravtsov, Yu.A., and A.I. Saichev. Effects of Double Passage of Waves in Random Inhomogeneous Media. *Sov. Phys. Usp.* 25 (1983): 494–508.
8. *Radar Handbook*. Ed. M.I. Skolnik. New York: McGraw- Hill, 1990.
9. Bertolotti, M. Photon Statistics. In Photon Correlation and Light Beating Spectroscopy. Eds. H.Z. Cummins and E.R. Pike. New York: Plenum, 1974.
10. O'Donnell, K.A. Speckle Statistics of Doubly Scattered Light. *J. Opt. Soc. Am.* 72 (1982): 1459–63.
11. Minnaert, M.G.J. *Light and Colour in the Outdoors*. New York: Springer-Verlag, 1993.
12. Oetking, P. Photometric Studies of Diffusely Reflecting Surfaces with Applications to the Brightness of the Moon. *J. Geophys. Res.* 71 (1966): 2505–13.
13. Dunbar, C. Reflection Factors of Photographic Papers. *Trans. Opt. Soc.* (London) 32 (1931): 184–96.
14. Bergman, G. Weak Localization in Thin Films. *Phys. Rev.* 107 (1984): 1–58.
15. *Scattering in Volumes and Surfaces*. Eds. M. Nieto-Vesperinas and J.C. Dainty. Amsterdam: North-Holland, 1990.
16. Jakeman, E., P.R. Tapster, and A.R. Weeks. Enhanced Backscattering through a Deep Random Phase Screen. *J. Phys. D: Appl. Phys.* 21 (1988): S32–S36.
17. Jakeman, E. Enhance Backscattering through a Deep Random Phase Screen. *J. Opt. Soc. Am.* A5 (1988): 1638–48.
18. Abramowitz, M., and I.A. Stegun. *Handbook of Mathematical Functions*. New York: Dover, 1971.
19. Tapster, P.R., A.R. Weeks, and E. Jakeman. Observation of Backscattering Enhancement through Atmospheric Phase Screens. *J. Opt. Soc. Am.* A6 (1989): 517–22.
20. Balmer, G.J., D.L. Jordan, P.R. Tapster, M.J. Kent, and E. Jakeman. Double Passage Wave Propagation Effects at Infra Red and Visible Frequencies. *Opt. Commun.* 88 (1992): 6–12.

21. Ward, K.D., and S. Watts. Radar Sea Clutter. In *Microwave J.* (1985): 109–21.
22. Pitter, M., E. Jakeman, and M. Harris. Heterodyne Detection of Enhanced Backscatter. *Opt. Letts.* 22 (1997): 393–95.
23. Jakeman, E., J.P. Frank, and G.J. Balmer. The Effect of Enhanced Backscattering on Target Detection. *AGARD Conf. Proc.* 542 (1993): 32/1–7.
24. Lilienfeld, P. Optical Detection of Particle Contamination on Surfaces: A Review. *Aerosol Sci. Technol.* 5 (1986): 145–65.

11

Vector Scattering: Polarization Fluctuations

11.1 Introduction

The foregoing chapters have been concerned with physical models for the scattering of *scalar* waves. These also provide a perfectly adequate description of many configurations involving the scattering and detection of a single polarized component of electromagnetic radiation, including both scattering by spherical particles, propagation through weakly scattering random media, and near-specular rough surface scattering when the tangent plane (phase screen) approximation can be used. However, in the case of scattering by moving anisotropic objects (for example, light scattering by nonspherical particles in suspension or grazing incidence backscattering from moving rough surfaces), the polarization state will change with time and this will be manifest in correlations between intensity fluctuations in different polarized components of the scattered radiation. It is well known that the state of polarization of a scattered electromagnetic wave contains additional information about the scattering object and, furthermore, can often be used as a *discriminant*, that is, exploited to distinguish a signal of interest from unwanted noise. Therefore, in this chapter some scattering models that take account of polarization effects will be described.

After a discussion of the polarization characteristics of Gaussian speckle, the case of scattering by small, freely diffusing spheroidal particles will be described in terms of the polarization states of the incident radiation and that selected by the detection system [1]. The model will then be extended to include the presence of a nearby surface. This is a configuration, mentioned at the end of the previous chapter, which has attracted attention in the context of optical monitoring of surface contamination in the semiconductor industry [2]. However, it is also important in polarization-sensitive geometries, such as low elevation-angle backscattering from rough surfaces that is relevant to many radar systems [3]. Indeed, differences in the fluctuations of horizontally and vertically polarized sea echos were noted over 50 years ago [4] and continue to exercise the maritime radar community (for a list of recent references see Reference [5]). Although results for average polarization cross sections will be given where appropriate, the main concern in this chapter will be with fluctuation effects, in line with the principal theme of the book.

11.2 The Polarization Characteristics of Gaussian Speckle

The simple random walk model described in Chapter 4 (Equation 4.6)

$$E(\bar{r},t) = \sum_{j=1}^{N} a_j \exp(i\varphi_j) \tag{11.1}$$

can be generalized to the case of vector scattering by associating initial and final polarization states with the detected field and the scattering amplitudes of each particle. In what follows, this will be denoted by an additional subscript on E and a_j. When N is large, the scattered field can be resolved into *two* complex Gaussian processes so that, when the phases in Equation 11.1 are uniformly distributed, the measurement of a single linearly polarized component will again lead to intensity fluctuations that are governed by an exponential distribution. In other words, when electromagnetic radiation that has been scattered by a large number of particles is viewed through a linear polarizing element, a fully developed Gaussian speckle pattern will be observed, whatever the initial and final polarization states of the field. On the other hand, in the absence of a polarizing element, the intensity is that of the sum of two Gaussian fields that may be correlated and have unequal means and variances. The general case may be expressed in terms of the joint probability density for two orthogonal vector components of the field and their correlation coefficient using the results in Chapter 2 for Gaussian processes

$$P(E_s^r, E_s^i, E_p^r, E_p^i) = \frac{1}{\pi^2 d} \exp\left\{ -[j_{11}I_s + j_{22}I_p - 2\operatorname{Re}(j_{12}E_s^* E_p)] / d \right\} \tag{11.2}$$

Here, the superscripts i and r denote the real and imaginary parts of the field, $E = E^r + iE^i$, while the component of the electric vector vibrating in the plane of scattering and at right angles to it are denoted by the subscripts p and s, respectively. The quantity d is the determinant of the coherency matrix [6]

$$J = \begin{pmatrix} j_{11} & j_{12} \\ j_{21} & j_{22} \end{pmatrix} = \begin{pmatrix} \langle I_p \rangle & \langle E_s^{\bullet} E_p \rangle \\ \langle E_p^{\bullet} E_s \rangle & \langle I_s \rangle \end{pmatrix} \tag{11.3}$$

$$d = j_{11} j_{22} - j_{12} j_{21}$$

In the case of Gaussian field components, Equation 11.2 can be used to derive the probability density of any measurable statistical property of the radiation using the results given in Chapter 2 for multivariate correlated Gaussian processes. Many such results have appeared in the literature during the past few decades, but a comprehensive analysis of fluctuations in the parameters of the Stokes vector

$$\begin{pmatrix} I \\ Q \\ U \\ V \end{pmatrix} = \begin{pmatrix} I_p + I_s \\ I_p - I_s \\ E_p E_s^{\bullet} + E_p^{\bullet} E_s \\ i(E_p^{\bullet} E_s - E_p E_s^{\bullet}) \end{pmatrix} \tag{11.4}$$

has been published comparatively recently [7,8]. It is not difficult to confirm that Equation 11.2 predicts the expected results in special cases. For example, the joint distribution of the s and p intensities takes the familiar form of Equation 2.21 for a joint Gaussian process

$$P(I_s, I_p) = \frac{1}{d} \exp\left(-\frac{1}{d}\left[j_{11} I_s + j_{22} I_p\right]\right) I_0\left(\frac{2|j_{12}|}{d}\sqrt{I_s I_p}\right) \tag{11.5}$$

The probability density of the first element in the Stokes vector (Equation 11.4), $I = I_s + I_p$, that is measured in the absence of a polarizing element in the detection system is

$$P(I) = \frac{1}{\wp \langle I \rangle}\left[\exp\left(-\frac{2I}{(1+\wp)\langle I \rangle}\right) - \exp\left(-\frac{2I}{(1-\wp)\langle I \rangle}\right)\right] \tag{11.6}$$

Here $\wp$ is the degree of polarization defined by

$$\wp = \frac{1}{\langle I \rangle}\sqrt{\langle Q \rangle^2 + \langle U \rangle^2 + \langle V \rangle^2} \tag{11.7}$$

and if it is zero then Equation 11.6 reduces to

$$P(I) = \frac{I}{\langle I \rangle^2} \exp\left(-\frac{I}{\langle I \rangle}\right)$$

This distribution is just the chi-square distribution of Equation 3.12 with m = 4 (or gamma distribution from Equation 3.13 with α = 2) corresponding to the two real and two imaginary components of the vector field that are identical independent Gaussian variables in this case. The other three Stokes parameters are governed by distributions of the form

$$P(y) = \frac{1}{\left[\langle I \rangle^2 (1-\wp^2) + \langle y \rangle^2\right]^{1/2}} \exp\left(\frac{1}{2d}\left\{y\langle y \rangle - |y|\left[\langle I \rangle^2 (1-\wp^2) + \langle y \rangle^2\right]^{1/2}\right\}\right)$$

(11.8)

At each point in the scattered field the polarization can be described in terms of an ellipse with axes I_a, I_b, ellipticity e, and orientation ψ defined by

$$I_a = \tfrac{1}{2}\left[I_s + I_p + \sqrt{(I_s - I_p)^2 + 4I_s I_p \cos^2 \delta}\right]$$

$$I_b = \tfrac{1}{2}\left[I_s + I_p - \sqrt{(I_s - I_p)^2 + 4I_s I_p \cos^2 \delta}\right]$$

(11.9)

$$\tan 2\psi = \frac{2\sqrt{I_s I_p}\cos\delta}{I_s - I_p}, \quad e = \sqrt{\frac{I_a}{I_b}}$$

where $\delta = \arg E_p - \arg E_s$. The statistical properties of these variables can also be calculated from Equation 11.2. For example, the difference variable $\zeta = I_a - I_b$ is governed by a member of a class of so-called I-K distributions that generalize the K-distribution model (Equation 3.25) as discussed in Chapter 12

$$P(\zeta) = \frac{\zeta}{d} I_0\left(\frac{\zeta}{2d}\sqrt{(j_{11} - j_{22})^2 + 4|j_{12}|^2 \cos^2\beta}\right) K_0\left(\frac{\zeta}{2d}\sqrt{(j_{11} - j_{22})^2 - 4|j_{12}|^2 \sin^2\beta}\right)$$

Here $\beta = \arg j_{12}$. The probability densities of ellipticity and orientation can also be obtained analytically from Equation 11.2, but the results are fairly complicated and the interested reader is referred to the original paper [7].

11.3 Non-Gaussian Polarization Effects in Scattering by Particles

Deviation from Gaussian statistics is conveniently expressed in terms of a generalization of the relation in Equation 4.29. Thus from the model (Equation 11.1), an expression for the correlation between intensities that have been measured under different polarization conditions can be derived

$$\frac{\langle I_u I_v \rangle}{\langle I_u \rangle \langle I_v \rangle} = \frac{\langle N(N-1) \rangle}{\langle N \rangle^2} \left(1 + |g_{uv}|^2 \right) + \frac{f_{uv}}{\langle N \rangle} \tag{11.10}$$

where

$$f_{uv} = \frac{\langle |a_u|^2 |a_v|^2 \rangle}{\langle |a_u|^2 \rangle \langle |a_v|^2 \rangle}, \qquad |g_{uv}|^2 = \frac{|\langle a_u a_v \rangle|^2}{\langle |a_u|^2 \rangle \langle |a_v|^2 \rangle} \tag{11.11}$$

It has been assumed that the scattering particles are statistically identical and the subscripts in these two equations now denote the initial to final polarization states. For example $u,v \equiv ss, pp, sp, ps$ where, for example, sp signifies s-polarized initial and p polarized final states.

The simplest scattering system that gives nontrivial polarization fluctuations is an assembly of small spheroidal particles. If the particles are sufficiently small compared to the wavelength, then a Rayleigh scattering or dipole approximation can be used to evaluate the statistical averages appearing in Equation 11.11. Referring to Figure 11.1, it will be assumed that each particle's axis of symmetry is aligned along the direction $\hat{n}_j = (\sin \psi_j \cos \phi_j, \sin \psi_j \sin \phi_j, \cos \psi_j)$. The field scattered into the detector by a single particle is given by [1]

$$a \propto \hat{e}.(\alpha \bar{E})$$

where $\bar{E}$ is the incident field, α is the polarizability tensor in the coordinate frame defined by $\hat{n}$, with

$$\alpha = \begin{pmatrix} \alpha_1 & 0 & 0 \\ 0 & \alpha_2 & 0 \\ 0 & 0 & \alpha_2 \end{pmatrix} \tag{11.12}$$

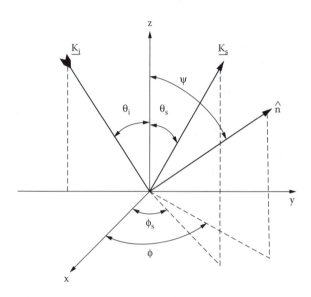

FIGURE 11.1
Scattering geometry.

and $\hat{e}$ projects out the measured polarization state. The usual convention for labelling these states with respect to the plane of scattering will be adopted as indicated above. For example $\hat{e}_s = (1, 0, 0)$ $\hat{e}_p = (0, \cos \theta_s, - \sin \theta_s)$ (11.13) Here the scattering angle is defined by $\theta = \pi - \theta_s$.

The statistical averages appearing in Equation 11.11 can be evaluated from the explicit relationships valid for scattering by small spheroidal particles

$$a_{ss} = E_s \left[(\alpha_1 - \alpha_2) \sin^2 \psi \cos^2 \phi + \alpha_2 \right]$$

$$a_{pp} = -E_p \left[(\alpha_1 - \alpha_2) \sin \psi \sin \phi (\sin \psi \sin \phi \cos \theta + \cos \psi \sin \theta) + \alpha_2 \cos \theta \right]$$

$$a_{ps} = E_p (\alpha_1 - \alpha_2) \sin^2 \psi \sin \phi \cos \phi$$

$$a_{sp} = -E_s \sin \psi \cos \phi (\sin \psi \sin \phi \cos \theta + \cos \psi \sin \theta)$$

$$(11.14)$$

Here E_s and E_p are the incident field components, and constant phase and propagation factors have been omitted. For the purpose of illustration it will be assumed that the particles are composed of material characterized by a real dielectric constant. Generalization to complex values is straightforward, but algebraically tedious. For a homogeneous system of randomly tumbling spheroids

$$P(\phi)d\phi = \frac{d\phi}{2\pi} \qquad 0 \le \phi \le 2\pi$$

$$P(\psi)d\psi = \frac{\sin \psi d\psi}{2} \qquad 0 \le \psi \le \pi .$$

(11.15)

The probability density functions of the amplitudes in Equation 11.14 can be obtained most easily from the characteristic functions $\langle \exp(i\lambda a \rangle$ that can be calculated by direct integration over the distributions of Equation 11.15 [9]. For example, for $\alpha_1 > \alpha_2 > 0$

$$P(a_{ss}) = \begin{cases} \left[2E_s(\alpha_1 - \alpha_2)(a_{ss} - \alpha_2 E_s) \right]^{-1/2} & \alpha_2 E_s < a_{ss} < E_s(\alpha_1 + \alpha_2)/2 \\ 0 & otherwise \end{cases}$$

(11.16)

Under the same conditions

$$P(a_{ps}) = \frac{2/\pi \sqrt{E_p(\alpha_1 - \alpha_2)}}{\sqrt{E_p(\alpha_1 - \alpha_2) + 2|a_{ps}|}} K \left(\sqrt{\frac{E_p(\alpha_1 - \alpha_2) - 2|a_{ps}|}{E_p(\alpha_1 - \alpha_2) + 2|a_{ps}|}} \right) \quad for \ |a_{ps}| < E_p(\alpha_1 - \alpha_2)$$

$$= 0 \qquad otherwise$$

(11.17)

where K is a complete elliptic integral. Plots of these functions are shown in Figure 11.2.

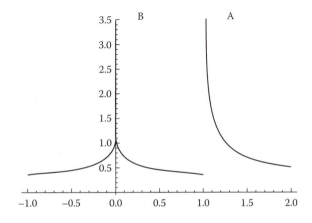

FIGURE 11.2
Amplitude probability distributions of freely tumbling particles, a) s-copolar, b) cross polar.

For the case of small particles, the parameters f and g appearing in Equation 11.10 are most easily evaluated by directly averaging powers and products of the amplitudes of Equation 11.14 over the distributions of Equation 11.15. It may readily shown that

$$\left\langle a_{ss}a_{sp}\right\rangle = \left\langle a_{ss}a_{ps}\right\rangle = \left\langle a_{pp}a_{ps}\right\rangle = \left\langle a_{pp}a_{sp}\right\rangle = 0 \tag{11.18}$$

so that most of the g values are zero. However, the non-Gaussian factors are generally nonzero. For example, setting $r = \alpha_1/\alpha_2$

$$f_{ss,ss} = \frac{\left\langle a_{ss}^4\right\rangle}{\left\langle a_{ss}^2\right\rangle^2} = \frac{5}{7}\frac{35r^4 + 40r^3 + 48r^2 + 64r + 128}{(3r^2 + 4r + 8)^2} \tag{11.19}$$

Results for a full range of polarization configurations are tabulated in Reference [9] and generally show a dependence on scattering angle as well as the polarization ratio r. Some examples of the non-Gaussian factors f are plotted in Figure 11.3, and it is clear that measurements of excess non-Gaussian fluctuations could be used to determine r and hence the shape of the particles provided that their refractive index was known. Non-Gaussian polarization fluctuations of this type can also be used in addition to the more familiar polarization cross sections to characterize and identify more complex radar and sonar targets. However, in the case of light scattering from particle suspensions, the method requires that only a few subwavelength-size particles be present in the illuminated volume and this will often be difficult to accomplish in practice. Moreover, the intensity of the scattered radiation may be small leading to unacceptably low signal to noise levels.

11.4 Correlation of Stokes Parameters in Particle Scattering

Although, according to Equation 11.18, in many polarization configurations the Gaussian parameters g are zero, $\langle a_{ss}\,a_{pp}\rangle$ and $\langle a_{sp}\,a_{ps}\rangle$ are nonvanishing so that there are evidently situations in which the Gaussian term in Equation 11.10 is finite and carries useful shape information. In the case of Rayleigh scattering, the following results are obtained

$$g_{ss,pp} = \frac{\left\langle a_{ss}a_{pp}\right\rangle^2}{\left\langle a_{ss}^2\right\rangle\left\langle a_{pp}^2\right\rangle} = \frac{\cos^2\theta\left[(r-1)^2 + 10(r-1) + 15\right]^2}{(3r^2 + 4r + 8)\left[(r-1)^2(3\cos^2\theta + \sin^2\theta) + 5(2r+1)\cos^2\theta\right]} \tag{11.20}$$

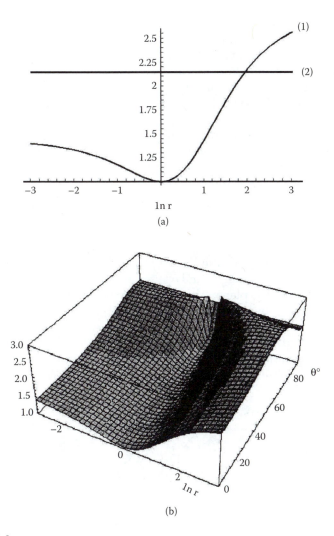

FIGURE 11.3
Fluctuation enhancement factors f versus polarization ratio and scattering angle a) s copolar
[1] and cross polar, all angles, and b) p copolar [9].

$$g_{sp,ps} = \frac{\left\langle a_{sp} a_{ps} \right\rangle^2}{\left\langle a_{sp}^2 \right\rangle \left\langle a_{ps}^2 \right\rangle} = \cos^2 \theta \qquad (11.21)$$

Since Gaussian speckle provides a much more robust source of information
than the deviation from Gaussian statistics expressed in the final term of
Equation 11.10, a good deal of effort has been devoted to the development
of techniques for the determination of particle shape based on measurements
of g. This work has generally been couched in terms of modified Stokes

parameters in which I_p and I_s are used instead of the first and second parameters in the definition in Equation 11.4.

The correlation functions of the Stokes parameters satisfy a relationship that is structurally identical to the generalized Siegert relation (Equation 11.10). In particular, the g (Gaussian) factors corresponding to the correlation of I with Q and to the second moments of U and V carry shape information [10]. These can be calculated for Rayleigh scattering using the approach outlined above from Equation 11.14. In practice, the finite size of real particles is taken into account by numerical simulation using a T-matrix technique to calculate the required scattering amplitudes [11]. Calculations show that fluctuations in the polarization state of the radiation scattered by a spheroid depend on only its aspect ratio for equivalent sphere radii that are less than about one tenth of the wavelength of the illuminating radiation (see Figure 11.4).

A diagram of apparatus constructed to shape particles in suspension at optical frequencies is shown in Figure 11.5 [12]. The measurement is based on correlation of the scattered s and p intensities with incident light polarized at 45° to the scattering plane. Although many other configurations are possible it has been shown that this one, together with incident right circular and correlation of right with left circular scattered intensities, is optimum. However, in order to achieve the most accurate measurement it is also necessary to optimize the scattering angle. Experimental results are compared with theory for haematite particles in Figure 11.6.

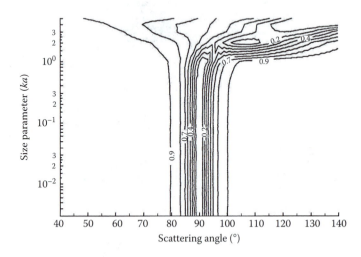

FIGURE 11.4
Contour plot of the equivalent volume spherical size parameter ka versus scattering angle for an oblate spheroid with aspect ratio 2.0 and complex refractive index $1.33 - i0.005$; contours display the value of the correlation coefficient $\langle I_p I_s\rangle / \langle I_p\rangle\langle I_s\rangle - 1$ [11].

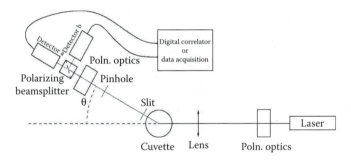

FIGURE 11.5
Diagram of apparatus for measuring particle shape using polarization fluctuation spectroscopy [12].

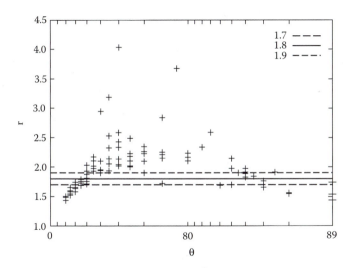

FIGURE 11.6
Recovered aspect ratios of haematite particles together with the nominal value of 1.8 ± 0.1 [12].

11.5 Scattering from Particles Near an Interface

Another scattering geometry that is commonly encountered in remote sensing systems is where the object of interest lies close to the bounding surface
between two materials. This configuration was discussed briefly at the end
of the previous chapter, but the polarization characteristics of the scattering
are of most interest. The model is obviously relevant to many radar and light
scattering scenarios, but it should be mentioned that the concept of droplets
or specular reflectors above the local surface has often been invoked to

explain the polarization characteristics of microwave sea echo [4] (a review of the literature in this area is given in Reference [5]).

In order to gain some insight into the problem, the simplest arrangement to study is that of a small spheroid above a perfectly flat interface (Figure 10.10). The treatment given here is not exact since it neglects perturbation of the surface field by the particle and the results for small particle-surface separations must therefore be treated with caution. However, results in the far field reduce to the correct solutions when the particle is actually on the surface[13] and numerical results appear to differ by less than 10% overall [14].

The scattered field is composed of four contributions as in the scalar case, but now the polarization states of the incident and reflected components must be included through appropriate use of the scattering coefficients (Equation 11.14) and the surface Fresnel reflection coefficients. The total return from a single particle is thus

$$
\begin{aligned}
s_u = {} & a_u(\theta_i, \theta_s) + R(\theta_i)a_u(\pi - \theta_i, \theta_s)\exp\left[if(\theta_i)\right] + \\
& + R(\theta_s)a_u(\theta_i, \pi - \theta_s)\exp\left[if(\theta_s)\right] + \\
& + R(\theta_i)R(\theta_s)a_u(\pi - \theta_i, \pi - \theta_s)\exp\left[if(\theta_i) + if(\theta_s)\right]
\end{aligned}
\tag{11.22}
$$

Here, the exponential factors take account of the path differences relative to the directly scattered contribution, with d being the height of the dipole above the interface

$$
f(\theta) = 2kd\cos\theta
\tag{11.23}
$$

The subscript u (= ss, pp, sp, ps) on the scattering amplitudes in Equation 11.22 can be referred to the surface so that s and p in Equation 11.14 are now identified as the horizontal H and vertical V components of the field, respectively. For clarity, this notation will be adopted for the remainder of the chapter. The complex field (Fresnel) reflection coefficients are defined in the usual way by

$$
R_H(\theta) = \frac{\cos\theta - \sqrt{\varepsilon - \sin^2\theta}}{\cos\theta + \sqrt{\varepsilon - \sin^2\theta}}, \quad R_V(\theta) = \frac{\varepsilon\cos\theta - \sqrt{\varepsilon - \sin^2\theta}}{\varepsilon\cos\theta + \sqrt{\varepsilon - \sin^2\theta}}
\tag{11.24}
$$

The result (Equation 11.22) can be used to calculate properties of the intensity fluctuations that arise due to spheroid reorientation, height variation, and particle number fluctuations.

11.6 Polarization Fluctuations: Particles Near an Interface

For the purpose of illustration, the discussion here will be restricted to Rayleigh scatterers in a backscattering configuration that is commonly encountered in longer wavelength remote sensing systems. The particle scattering geometry is illustrated in Figure 11.1 so that from relations of Equation 11.14 the H copolar scattered field component is given by [5]

$$s_{HH} = E_H \left[(r-1)\sin^2 \psi \cos^2 \varphi + 1 \right] \left[1 + R_H \exp(if) \right]^2 \tag{11.25}$$

In this case the fluctuations due to particle reorientation and those due to particle height variations above the surface factorize so that averaging over realizations of particle orientation using the distributions of Equation 11.15 leads to

$$\left\langle \left| s_{HH} \right|^2 \right\rangle = \left| E_H \right|^2 \left[1 + \tfrac{2}{3}(r-1) + \tfrac{1}{5}(r-1)^2 \right] \left\langle \left| 1 + R_H \exp(if) \right|^4 \right\rangle \tag{11.26}$$

The average in the last factor of the right-hand side of this result can be evaluated for the appropriate choice of distribution of the height d appearing in Equation 11.23. For example, in the case of a Gaussian distribution with rms height d_0 and a perfectly conducting surface, $R_H = -1$, the following result is obtained

$$\left\langle \left| 1 + R_H \exp(if) \right|^4 \right\rangle = 2(3 - 4x + x^4)$$

$$x = \exp(-2k^2 d_0^2 \cos^2 \theta) \tag{11.27}$$

The dependence of the result in Equation 11.26 on the polarizability ratio r of the particles shows that, as might be expected, a rod-like or prolate spheroidal particle ($r > 1$) scatters more strongly than an oblate one. The cross section falls to zero near grazing incidence and also as the particle-to-surface distance is reduced.

The normalized second moment of the fluctuations in the H copolar return from a particle above an interface can also be calculated without difficulty. After averaging over fluctuations in orientation, it may be expressed in the factored form

$$\frac{\left\langle\left|s_{HH}\right|^4\right\rangle}{\left\langle\left|s_{HH}\right|^2\right\rangle^2} = \frac{1+\frac{4}{3}(r-1)+\frac{6}{5}(r-1)^2+\frac{4}{7}(r-1)^3+\frac{1}{9}(r-1)^4}{\left[1+\frac{2}{3}(r-1)+\frac{1}{5}(r-1)^2\right]^2} \frac{\left\langle\left|1+R_H\exp(if)\right|^8\right\rangle}{\left\langle\left|1+R_H\exp(if)\right|^4\right\rangle^2}$$

(11.28)

For a fixed particle height the largest effect occurs for prolate spheroids but the maximum value of 25/9 is by no means extreme. In the case of a Gaussian distribution of height and a perfectly conducting surface

$$\left\langle\left|1+R_H\exp(if)\right|^8\right\rangle = 2(35-56x+28x^4-8x^9+x^{16})$$ (11.29)

This result may be combined with Equation 11.27 to give the last factor in Equation 11.28. It is evident from Figure 11.7 for a sphere that the fluctuations arising from changes of height can be exceptionally large near grazing where the second normalized moment approaches 35/3. If the root square height is small compared to a wavelength, the region of large fluctuation extends to higher elevation angles; however, the theory is not strictly accurate in this regime. The horizontal copolar results for a particle above a dielectric are not dissimilar to those for the perfectly conducting case.

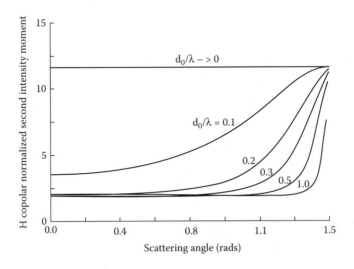

FIGURE 11.7
H copolar backscattered intensity fluctuations due to Gaussian height variations (rms d_0) of a sphere above a perfectly conducting surface; plot of second normalized intensity moment against scattering angle [5].

Results for the vertical copolar cross section of scattering by a single particle above an interface have a more complicated dependence on angle and the fluctuations due to reorientation of the particle, and changes in its height above the interface are not separable as in the case of horizontally polarized scattering. Full results for this case are given in Reference [5]. The angle dependence of the vertical copolar backscattering cross section is relatively insensitive to the polarizability ratio r. The scattered intensity is largest near grazing incidence due to the enhanced coupling (particularly at small heights) between the field and the dumbbell scatterer composed of the particle and its image in the surface. As in the case of H copolar scattering, the fluctuations in the return, due to particle reorientation, are relatively modest, but those due to height variations may be larger. This effect is most marked for dielectric surfaces near grazing incidence. Unlike the horizontally polarized results, the fluctuations remain small at all angles of incidence for a particle above a highly conducting surface.

In the case of scattering by many particles above an interface, a generalized Siegert relation of the form given in Equation 11.10 applies with the functions f and g now defined in terms of the statistical properties of s, such as results in Equation 11.26 through Equation 11.29. However, a configuration that is often of more interest is that of a single scatterer above a *rough* surface as discussed in Section 10.6.

References

1. Van de Hulst, H.C. *Light Scattering by Small Particles.* New York: Wiley, 1957.
2. Lilienfeld, P. Optical Detection of Particle Contamination on Surfaces: A Review. *Aerosol Sci. Technol.* 5 (1986): 145–65.
3. Skolnik, M.I., ed. *Radar Handbook.* 2nd Edition. New York: McGraw-Hill, 1990.
4. Goldstein, H. The Fluctuations in Sea Echo. In *Propagation of Short Radio Waves.* Ed. D.E. Kerr. New York: Dover, 1965.
5. Jakeman, E., D.L. Jordan, and G.D. Lewis. Fluctuations in Radiation Scattered by Small Particles Above an Interface. *Waves in Random Media* 10 (2000): 317–36.
6. Goodman, J.W. *Statistical Optics.* New York: Wiley, 1985.
7. Eliyahu, D. Vector Statistics of Gaussian Fields. *Phys. Rev. E* 47 (1993): 2881–92.
8. Mendez, E.R., A.G. Navarrete, and R.E. Luna. Statistics of the Polarization Properties of One-Dimensional Randomly Rough Surfaces. *J. Opt. Soc. Am.* A12 (1995): 2507–16.
9. Jakeman, E. Polarisation Characteristics of Non-Gaussian Scattering by Small Particles. *Waves in random media* 5 (1995): 427–42.
10. Bates, A.P., K.I. Hopcraft, and E. Jakeman. Non-Gaussian Fluctuations of Stokes Parameters in Scattering by Small Particles. *Waves in Random Media* 8 (1998): 235–53.
11. Pitter, M.C., K.I. Hopcraft, E. Jakeman, and J.G. Walker. Structure of Polarisation Fluctuations and Their Relation to Particle Shape. *J. Quant. Spectr. and Rad. Trans.* 63 (1999): 433–44.

12. Chang, P.C.Y., K.I. Hopcraft, E. Jakeman, and J.G. Walker. Optimum Configuration for Polarisation Photon Correlation Spectroscopy. *Meas. Sci. Technol.* 13 (2002): 341–48.
13. Jakeman, E. Scattering by Particles on an Interface. *J. Phys. D: Appl. Phys.* 27 (1994): 198–210.
14. Videen, G., W.L. Wolfe, and W.S. Bickel. Light Scattering Mueller Matrix for a Surface Contaminated by a Single Particle in the Rayleigh Limit. *Opt. Eng.* 31 (1992): 341–49.

12

K-Distributed Noise

12.1 Introduction

In preceding chapters the class of non-Gaussian K-distributions has often been encountered. In Chapter 3 these arose as the distribution of the gradient of a signal that was gamma distributed (Equation 3.23) and also as a compound process in which the average intensity of a complex Gaussian variable was itself a gamma distributed variable (Equation 3.24). In Chapter 4 it was shown that in the limit of a large *average* number of steps the resultant of a random walk in which the number of steps varied according to a negative binomial distribution was a K-distributed variable (Equation 4.28). It was shown in Chapter 7 (Equation 7.19) that fluctuations in the intensity of radiation that has been scattered by a Brownian subfractal phase screen are governed by a K-distribution, while in Chapter 10 (Equation 10.1 and Equation 10.9) it was demonstrated that certain multiple scattering scenarios lead to this kind of intensity fluctuation. With the exception of the first example, these mechanisms are all compound *variables*. K-distributed noise is defined to be a time/space dependent compound *process* that generates a variable whose marginal probability density is a K-distribution [1]. Many phenomena can be characterized as the modulation of one process by another and so it is perhaps not surprising that as a consequence the class of K-distributions has found many applications in science, engineering, and more recently in medicine. In this chapter the properties of this useful non-Gaussian model and its developments and generalizations will be reviewed.

From a mathematical point of view the simplest mechanism leading to K-distributed noise is a Gaussian process whose mean square is modulated by a gamma variate [2]. Note that gamma distributions are the continuum analog of the negative binomial class of discrete distributions that were used in the derivation of K-distributions given in Chapter 4. Physically, this compound process could correspond to a Gaussian speckle pattern with a "locally" varying mean intensity arising from density fluctuations within a large population of scattering centers as described in Chapter 4. The local changes in mean intensity might be visible in an optical scattering pattern, for example, as spatial variations in brightness of the speckle pattern, or be manifest as intermittent brightening of the whole pattern due to overall changes of scatterer density in the illuminated volume. In either case, the signal generally fluctuates on more than one time scale and the corresponding spatial pattern is characterized by more than one length. The shortest time scale in the case of short wavelength transmissions is due to interference between the returns from different scatterers or independent elements of the target, that is, the Gaussian or speckle contribution to the scattered radiation (see Equation 4.30), while longer time variations will be associated with the scatterer density fluctuation time. Similarly, the smallest length scale is the speckle size, that is, the diffraction lobe-width dictated by the illuminated volume, while the longer scales will relate to the spatial extent of scatterer density fluctuations.

12.2 Experimental Evidence

Although *K*-distributions had previously figured on occasion in the radio physics [3] and population statistics [4] literature, their importance in characterizing the fluctuation statistics of microwave returns from the sea surface was first recognized in 1976 [2,5]. The original proposal was based on two desirable features of *K*-distributions: (a) their fluctuation characteristics, measured by their normalized moments, lie between Rayleigh and lognormal as observed experimentally, and (b) they are consistent with a finite two-dimensional random walk model for scattering, changing analytically with scatterer number (or illuminated area) as expected on physical grounds. Feature (b) embodies a useful scaling property of *K*-distributions that can be demonstrated as follows. As seen in Chapter 3 (Equation 3.25), *K*-distributions are defined in terms of modified Bessel functions of the second kind, *K* (Figure 12.1)

$$P(I) = \frac{2b}{\Gamma(\alpha)}(bI)^{\frac{\alpha-1}{2}} K_{\alpha-1}\left(2\sqrt{bI}\right) \tag{12.1}$$

The characteristic function, corresponding to a sum of N independent, statistically identical vectors whose amplitudes, a, $(I = a^2)$ are governed by this distribution, is given by

$$\left\langle \exp\left[iu\sum_{j=1}^{N} a_j \exp i\varphi_j\right]\right\rangle = \left\langle J_0(ua)\right\rangle^N = \left(1 + u^2\left\langle a^2\right\rangle\big/4N\alpha\right)^{-N\alpha} \tag{12.2}$$

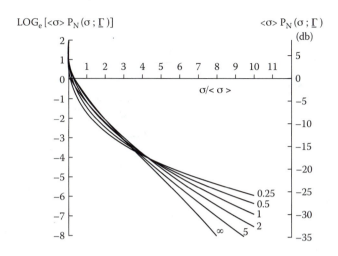

LOG$_e$ [<σ> P$_N$(σ; Γ)]   <σ> P$_N$(σ; Γ) (db)

σ/< σ >

FIGURE 12.1

K-distributions for various values of the index; log-linear plot shows negative exponential distribution as a straight line as α → ∞ [5].

Inverse Bessel transformation leads back to Equation (12.1) but with α replaced by Nα (see Equation 4.28). Thus the sum of a finite number, N, of independent vectors whose amplitudes a are K-distributed are also K-distributed but with a scaled shape parameter α → Nα. The distributions are said to be *infinitely divisible*. This is a useful property in the statistical analysis of system performance. For example, the coherent addition of independent vectors is often used to model the effect of illuminating independently contributing elements of a scattering target. Since one might anticipate that the number of independent scatterers will be proportional to the total area illuminated for the case of an extended target, a plot of normalized second moment against (1/illuminated area) should be a straight line from the ordinate two at abscissa zero, which corresponds to the large illuminated area limit. This behavior has indeed been observed in light scattering experiments [6]. However, it does not appear to hold for microwave sea echo near grazing incidence, where it is found experimentally that the K-distribution shape parameter scales like resolution length to the power of 5/8 [7]. This is probably a consequence of shadowing and the multiscale nature of the sea surface, which requires a more sophisticated definition of "independent" contributing areas than used in the previous argument. It has recently been shown how the required range correlation of K-distributed microwave sea echo can be simulated numerically through memory-less nonlinear transformation of a fractal [8].

An excellent fit to the available microwave sea echo data was provided by K-distributions (Figure 12.2) and indeed a time series analysis confirms the existence of multiple time scales expected on the basis of the random walk model (Equation 4.29). Thus, the intensity autocorrelation function exhibits a short time decay associated with interference between returns from many

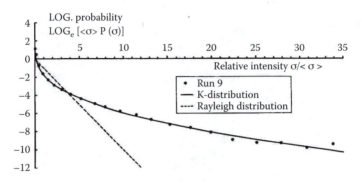

FIGURE 12.2
Experimental microwave sea echo data fitted to a K-distribution [15].

small scattering structures and a longer-term falloff related to the slow motion of individual larger waves.

The *K*-distributed noise model was subsequently found to fit fluctuation data obtained in other fields of science and engineering, particularly systems in which the scattering medium was characterized by several scales. These include microwave land clutter [9], light scattering from electrically driven turbulent convection in thin layers of nematic liquid crystal [10], laser propagation through thermal plumes and atmospheric turbulence [11], and ultrasound scattering from human tissue [12]. Normalized moments of fluctuations in radiation scattered by these systems were often compared with those of *K*-distributions

$$\frac{\langle I^n \rangle}{\langle I \rangle^n} = \frac{n!\,\Gamma(n+\alpha)}{\alpha^n \Gamma(\alpha)} \tag{12.3}$$

The higher moments can easily be calculated from Equation 12.3, but usually data is inadequate for comparison for $n > 5$. It is clear that if α is deduced from the second normalized moment then all the other moments can be predicted, that is, there is a unique relationship between the second and higher normalized moments for *K*-distributions which can be compared with experimental data. Figure 12.3 and Figure 12.4 show data compared with theoretical plots of this kind. The agreement appears to be excellent but some reservations have to be expressed about this mode of comparison since the skewed nature of the distributions causes downward bias of the moment data in any experiment of finite duration [13]. This problem will be discussed further in the next chapter.

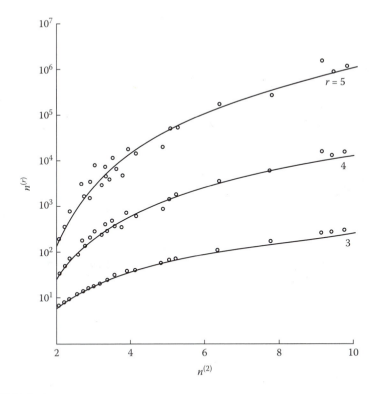

FIGURE 12.3
Liquid crystal scattering data compared with the moments of K-distributed noise $n^{[r]} = \langle I^{[r]} \rangle / \langle I \rangle^r$ [10].

12.3 A Population Model for Scatterer Number Fluctuations

Whereas Gaussian speckle can be readily attributed, through application of the central limit theorem, to interference between many randomly phased contributions to the received intensity, the origin of the underlying gamma process or scatterer density fluctuations required to generate K-distributed noise will be more problem specific. One model that has been extensively investigated in this context is the birth-death-immigration process of classical population statistics. This generates a negative binomial number of individuals (i.e., scattering centers) that is known to lead in the random walk model to K-distributions (Equation 4.26, Equation 4.27, and Equation 4.28) when the average number of steps is large. Imagine a population in which individuals are born and die at rates that are proportional to the existing number in the population and into which there is a steady stream of immigrants that is independent of this number. The time evolution of the probability $P_N(t)$

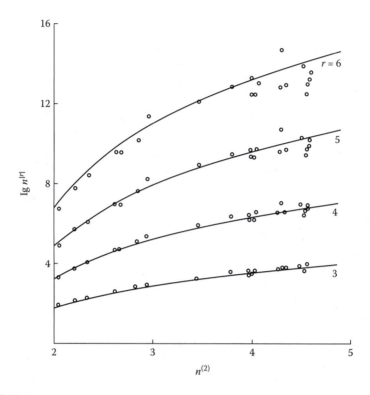

FIGURE 12.4
Data from various turbulence scattering experiments compared with the moments of K-distributed noise $n^{[r]} = \langle I^{[r]} \rangle / \langle I \rangle^r$ [15].

of finding a population of N individuals at time t is governed by an equation that balances these transition rates.

Figure 12.5 represents this process by a simple diagram of transitions between population levels and allows us to construct the following rate equation

$$\frac{dP_N}{dt} = \mu(N+1)P_{N+1} - \left[(\lambda+\mu)N + \nu\right]P_N + \left[\lambda(N-1) + \nu\right]P_{N-1} \quad (12.4)$$

This process has been studied for many years and the solutions of Equation 12.4 are well known [14]. The equation describes a first-order Markov process that may be solved for the conditional probability of finding a population of N individuals at time t_1 given that there was M present at an earlier time t_0. Solution is facilitated through the use of a generating function

$$Q(s;t) = \sum_{N=0}^{\infty}(1-s)^N P_N(t) \quad (12.5)$$

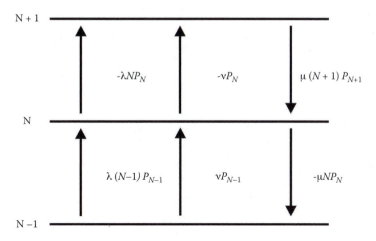

FIGURE 12.5
Transitions between population levels of a birth-death-immigration process.

From this the factorial moments and probabilities may be calculated by differentiation

$$n^{[r]} = \frac{\langle N(N-1)(N-2)\cdots(N-r+1)\rangle}{\langle N\rangle^r} = \left(-\frac{1}{\langle N\rangle}\frac{d}{ds}\right)^r Q(s;t)\Big|_{s=0}$$

(12.6)

$$P_N = \frac{1}{N!}\left(-\frac{d}{ds}\right)^N Q(s;t)\Big|_{s=1}$$

The equation for Q corresponding to (12.4) is

$$\frac{\partial Q}{\partial t} = s[\lambda(1-s)-\mu]\frac{\partial Q}{\partial s} - vsQ$$

(12.7)

After Laplace transformation with respect to t and incorporating the boundary condition at $t = 0$, a nonlinear first-order differential equation in s is obtained that can be solved in a straightforward manner. If there are M individuals present initially, this leads to the following well-known result [14]

$$Q^{(M)}(s,t) = \left(\frac{\mu-\lambda}{\mu-\lambda+\lambda(1-\theta)s}\right)^{v/\lambda}\left(\frac{\mu-\lambda+(\lambda-\mu\theta)s}{\mu-\lambda+\lambda(1-\theta)s}\right)^M$$

(12.8)

Here $\theta(t) = \exp[\lambda - \mu)t]$ and it is clear that an equilibrium distribution exists only if $\mu > \lambda$. When this inequality is satisfied, the excess deaths are balanced by immigrants and the following generating function for the equilibrium distribution is obtained

$$Q_E(s) = Q(s, \infty) = \left(1 + \frac{\lambda s}{\mu - \lambda}\right)^{-v/\lambda} \tag{12.9}$$

The corresponding equilibrium distribution is negative binomial and its normalized factorial moments can be constructed from this result using Equation 12.6 (cf. Equation 4.26)

$$P_N = \binom{N + v/\lambda - 1}{N} \frac{\left(\lambda \bar{N}/v\right)^N}{\left(1 + \lambda \bar{N}/v\right)^{N + v/\lambda}} \tag{12.10}$$

$$\bar{N} = v/(\mu - \lambda) \tag{12.11}$$

$$n^{[r]} = \frac{\Gamma\left(r + v/\lambda\right)}{\left(v/\lambda\right)^r \Gamma\left(v/\lambda\right)} \tag{12.12}$$

The second normalized factorial moment is of particular interest and takes the form

$$n^{[2]} = 1 + \lambda/v \tag{12.13}$$

Because of the Markov nature of the process all of its higher-order joint correlation properties can be constructed from the conditional solution of Equation 12.8. Thus

$$P\left(M(0), N(t)\right) = P(M, N) = P_M P(N \mid M) \tag{12.14}$$

Here P_M is the equilibrium distribution given by Equation 12.9 and $P(N \mid M)$ is the conditional distribution obtained from Equation 12.8. The correlation function or bilinear moment is thus given by

$$\frac{\langle MN \rangle}{\bar{N}^2} = -\frac{1}{\bar{N}^2} \sum_{M=0}^{\infty} M P_M \left. \frac{\partial Q^{(M)}(s, t)}{\partial s} \right|_{s=0} = 1 + \left(\frac{\lambda}{v} + \frac{1}{\bar{N}}\right)\theta \tag{12.15}$$

Notice that this does not reduce to Equation 12.13 when the delay time is zero since Equation 12.15 is then the normalized mean square population number, not the factorial moment.

12.4 Properties of the Scattered Intensity

The results shown above may be used to calculate the normalized intensity correlation function for the random walk model (Equation 4.6) [15]. The result may be written down exactly in the form

$$g^{(2)}(\tau) = 1 + \left(1 + \frac{1}{\alpha}\right)\left|g^{(1)}(\tau)\right|^2 + \left(\frac{1}{\alpha} + \frac{1}{\overline{N}}\right)\theta(\tau) + \frac{1}{\overline{N}}\left(\frac{\left\langle a^2(0)a^2(\tau)\right\rangle}{\left\langle a^2\right\rangle^2} - 1\right)\exp(-\mu\tau)$$

(12.16)

Here the first-order correlation function is given by

$$g^{(1)}(\tau) = \exp(-\mu\tau)\left\langle a(0)a(\tau)\right\rangle\left\langle \exp\left[i\left(\varphi(0) - \varphi(\tau)\right)\right]\right\rangle$$

(12.17)

$\overline{N}$ is given by Equation 12.11, $\theta(\tau) = \exp[(\lambda - \mu)\tau]$ as before and $\alpha = \nu/\lambda$. In the high density limit this reduces to

$$g^{(2)}(\tau) = 1 + \left(1 + \frac{1}{\alpha}\right)\left|g^{(1)}(\tau)\right|^2 + \frac{1}{\alpha}\theta(\tau)$$

(12.18)

Equation 12.18 reduces to the Siegert (Gaussian) factorization theorem (Equation 2.25) when $\alpha \to \infty$ and the clustering or number density fluctuation vanishes. In the limit of small delay time the result approaches the second normalized moment of the K-distribution given by Equation 12.3 with $n = 2$

$$\frac{\left\langle I^2\right\rangle}{\left\langle I\right\rangle^2} = 2 + \frac{2}{\alpha}$$

(12.19)

In the case of short wave scattering, the phase fluctuation term in Equation 12.17, due to interference between returns from different scatterers, will be characterized by a shorter time scale than the scatterer density fluctuations. The first-order correlation function will therefore generally be dominated by the phase fluctuation factor while the second-order correlation function will exhibit two distinct timescales: one associated with interference and the other with density fluctuations. These two contributions can readily be identified in the result of Equation 12.18 and are often observed in non-Gaussian scattering experiments as indicated in Section 12.3. Note that when the interference term is neglected, Equation 12.18 reduces to the high density limit of the number fluctuation result of Equation 12.15.

In principle, the higher-order factorization properties of K-distributed noise based on the above population model can be evaluated, but the calculations are complicated and to date no further results have appeared in the literature. Note that because the population model is Markovian, the spectrum of the number fluctuation component is restricted to be Lorentzian (corresponding to a negative exponential correlation function). It has been suggested that simulation of K-distributed noise with spectral behavior more accurately matching experimental data can be achieved by using the method of memory-less nonlinear transformation [8]. This method will be discussed in Section 14.6 but it should be emphasized that, unlike the case of Gaussian noise where the marginal distribution and spectrum are sufficient to entirely characterize the process, the higher-order noise properties of K-distributed noise such as triple correlations will not necessarily be generated correctly in this way.

12.5 Related Distributions

Various generalizations of the K-distribution model have been proposed to deal with particular applications. The simplest extension, in principle, is to add a coherent or unscattered contribution of magnitude a_0 to a randomly phased vector with K-distributed amplitude [15]. Unfortunately, the resultant *homodyned* K-distributed noise that is obtained by averaging the noise mean of a Rice distribution (I_G in Equation 3.5) over a gamma distribution (Equation 3.13) cannot be written in a simple closed form apart from the special case $\alpha = 1$ which gives the result

$$
\begin{aligned}
P(I) &= 2bI_0\left(2\sqrt{bI}\right)K_0\left(2a_0\sqrt{b}\right) \qquad 0 < I < a_0^2 \\
&= 2bI_0\left(2a_0\sqrt{b}\right)K_0\left(2\sqrt{bI}\right) \qquad a_0^2 < I < \infty
\end{aligned}
\tag{12.20}
$$

On the other hand, the intensity moments can quite generally be represented as finite sums

$$\left\langle I^r \right\rangle = \frac{(r!)^2}{\Gamma(\alpha)} \left(\frac{1^r}{b} \right)^r \sum_{p=0}^{r} \frac{\Gamma(r-p+\alpha)}{(p!)^2(r-p)!} \left(a_0^2 b \right)^p \qquad (12.21)$$

This result has been compared with optical data acquired in experiments where light is scattered by a turbulent layer [15] and more recently utilized in tissue characterization by ultrasonic techniques [16]. It is perhaps worth noting that when $\alpha = 1$ result of Equation 12.20 can be generalized in a different way by averaging the sum of a constant component and a Gaussian process in n dimensions over fluctuations in the mean of the noise component. This leads to the so-called *I-K* class of distributions

$$P(I) = \left(\frac{nba_0}{\sqrt{I}} \right) \left(\frac{\sqrt{I}}{a_0} \right)^{n/2} \times \begin{cases} I_{n/2-1}\left(\sqrt{2nbI} \right) K_{n/2-1}\left(a_0\sqrt{2nb} \right) & 0 < I < a_0^2 \\ I_{n/2-1}\left(a_0\sqrt{2nb} \right) K_{n/2-1}\left(\sqrt{2nbI} \right) & a_0^2 < I < \infty \end{cases}$$

$$(12.22)$$

These distributions have been used to model the fluctuations in laser radiation that has propagated through atmospheric turbulence [17] although the physical basis for this is open to question.

A more tractable extension of the model is based on the idea of a *biased* two-dimensional random walk in which the direction of the individual steps is not isotropic. The *generalized K-distribution* is obtained in the large scatterer density limit when the step number is clustered according to a negative binomial distribution. This gives [18]

$$P(I) = \frac{2\sqrt{b'/I}}{\Gamma(\alpha)} \left[\sqrt{b'I} \left(\frac{b}{b'} \right) \right]^\alpha I_0(2\sqrt{cI}) K_{\alpha-1}(2\sqrt{b'I}) \qquad (12.23)$$

where $b'^2 = b^2 + c^2$ and c is a parameter which measures the degree of bias in the walk. When c is zero Equation 12.23 reduces to Equation 12.1, and when α is large it approaches a Rice distribution (Equation 3.5). A further generalization of the model can again be based on random walks in higher-dimensional spaces but is not generally relevant to the wave scattering problems of interest here. Although originally proposed in the context of optical propagation through the atmosphere [19], Equation 12.23 has found more appropriate applications in the characterization of polarimetric and interferometric synthetic aperture radar data [20].

12.6 Statistical Mechanics

The K-distributed noise model has been successfully applied to the evaluation of clutter limited radar system performance, and has resulted in the development of new and more effective signal processing techniques for target detection. In such applications, it is often useful to be able to numerically simulate signal and noise data because the acquisition of real data can be prohibitively expensive. This requirement has led to the development of a statistical mechanics for K-distributed noise along the lines of Gaussian noise theory [21]. A rigorous treatment of this problem has been given recently by Field and Tough [22], and only the simplest exposition is given here [1].

In the high density limit, it can be shown that the rate Equation 12.4 becomes the rate equation for the conditional distribution of a gamma-Markov process

$$\frac{\partial P}{\partial t} = \lambda I \frac{\partial^2 P}{\partial I^2} + \left[2\lambda - \nu + (\mu - \lambda)I\right]\frac{\partial P}{\partial I} + (\mu - \lambda)P \tag{12.24}$$

Note that the Laplace transform of this equation with respect to I is identical to Equation 12.7 since the negative binomial distribution can be represented as a doubly stochastic Poisson process, that is,

$$P_N = \frac{1}{N!}\int_0^\infty dI \, P(I)I^N \exp(-I) \tag{12.25}$$

from which it follows that $\langle(1 - s)^N\rangle = \langle\exp(-sI)\rangle$. Making the substitution $I = A^2$, $P(I, t) = A^{-1} R(A, t)$ in Equation 12.24 obtains an equation for the conditional distribution of the generalized Rayleigh-Markov process

$$\frac{\partial R}{\partial t} = \frac{\partial}{\partial A}\left\{\left[\frac{\lambda - 2\nu}{4A} + \frac{(\mu - \lambda)A}{2}\right]R\right\} + \frac{\lambda}{4}\frac{\partial^2 R}{\partial A^2} \tag{12.26}$$

This has the standard form of a Fokker-Planck equation and corresponds to the Langevin equation, or equation of motion

$$\frac{dA}{dt} = \frac{2\nu - \lambda}{4A} - \frac{(\mu - \lambda)A}{2} + f\sqrt{\frac{\lambda}{2}} \tag{12.27}$$

where f is delta-correlated Gaussian noise. The relation of Equation 12.27 can be used to generate a signal that is gamma distributed with a negative exponential correlation function, that is, gamma-Lorentzian noise. This approach can be extended to generate K-distributed noise by using the output from Equation 12.27 to drive the mean of a second gamma-Markov process that has $\alpha = 1$. This leads to a pair of coupled Langevin equations of the form

$$
\frac{dx}{dt} = a\left(\alpha + \frac{z}{x} - x\right) + \sqrt{2ax}f_1(t)
$$

$$
\frac{dz}{dt} = b\left(1 - \frac{z}{x}\right) + \sqrt{2bz}f_2(t)
$$

(12.28)

where a, b are constants and f_1, f_2 are delta-correlated Gaussian noises. These equations have been used to generate individual realizations of the process, and favorable comparisons have been made with measured data (Figure 12.6). A "compound" interpretation of K-distributed noise provides various

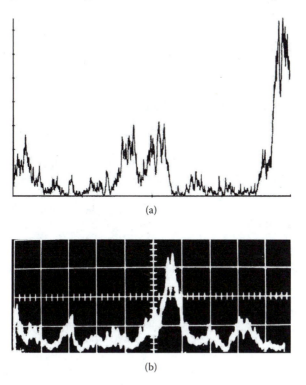

(a)

(b)

FIGURE 12.6
(a) Simulated K-distributed noise compared with (b) intensity fluctuations in light scattered by electrohydrodynamic turbulence in a nematic liquid crystal [1].

options for such simulation, however, given the power of modern computers [23].

A more general approach to the statistical mechanics underlying K-distributed Markov processes has recently been published [24]. It has already been pointed out that K-distributions can arise in a number of ways, and the new theory derives a more general pair of coupled stochastic differential equations and associated Fokker-Planck equations that lead to K-distributed noise. Signal processing techniques are proposed that can identify the particular mechanism responsible for K-distributed noise in given data.

Finally, we note that the topic of K-distributions and their application to radar performance, particularly in the presence of sea clutter, forms the subject of book that is in press at the time of writing. This provides an in-depth description of one of the most important practical applications of the model [25].

References

1. Jakeman, E., and R.J.A. Tough. Non-Gaussian Models for the Statistics of Scattered Waves. *Adv. Phys.* 37 (1988): 471–529.
2. Ward, K.D. Compound Representation of High Resolution Sea Clutter. *Electron. Lett.* 17 (1981): 561–65.
3. Siddiqui, M.M., and G.H. Weiss. Families of Distributions for Hourly Median Power and Instantaneous Power of Radio Signals. *J. Res. Nat. Bur. Stds.* 67D (1963): 753–62.
4. Norikazu, Y. The Random Walk Model of Human Migration. *Theor. Pop. Biol.* 7 (1975): 156–67.
5. Jakeman, E., and P.N. Pusey. A Model for Non-Rayleigh Sea Echo. *IEEE Trans. Antenn. Prop.* AP24 (1976): 806–14.
6. Jakeman, E., and P.N. Pusey. Non-Gaussian Fluctuations in Electromagnetic Radiation Scattered by a Random Phase Screen II: Application to Dynamic Scattering in a Liquid Crystal. *J. Phys. A.* 8 (1975): 392–410.
7. Ward, K.D., and S. Watts. Radar Sea Clutter. *Microwave J.* (June 1985): 109–21.
8. Ward, K.D., R.J.A. Tough, and P.W. Shepherd. Modelling Sea Clutter: Correlation, Resolution and Non-Gaussian Statistics. *Radar 97* (IEE Conference Publication 449, 1997): 95–99.
9. Oliver, C.J. A Model for Non-Rayleigh Scattering Statistics. In *Wave propagation and scattering.* Ed. B.J. Uscinski. Oxford: Clarendon, 1986, 155–73.
10. Jakeman, E., and P.N. Pusey. Significance of K-distributions in Scattering Experiments. *Phys. Rev. Lett.* 40 (1978): 546–48.
11. Parry, G., and P.N. Pusey. K-distributions in Atmospheric Propagation of Laser Light. *J. Opt. Soc. Am.* 69 (1979): 796–98.
12. Molthen, R.C., P.M. Shankar, and J.M. Reid. Characterisation of Ultrasonic B-Scans Using Non-Rayleigh Statistics. *Ultrasound in Med. & Biol.* 21 (1995): 161–70.

13. Tapster, P.R., A.R. Weeks, P.N. Pusey, and E. Jakeman. Analysis of Probability-Density Functions for Laser Scintillations in a Turbulent Atmosphere. *J. Opt. Soc. Am.* A6 (1989): 782–85.

14. Bartlett, M.S. *An Introduction to Stochastic Processes.* Cambridge: Cambridge University Press, 1966.

15. Jakeman, E. On the Statistics of K-Distributed Noise. *J. Phys. A: Math. Gen.* 13 (1980): 31–48.

16. Dutt, V., and J.F. Greenleaf. Ultrasound Echo Envelope Analysis Using a Homodyned *K*-Distribution Model. *Ultrasonic Imaging* 16 (1994): 265–87.

17. Andrews, L.C., and R.L. Phillips. I-K Distributions as a Universal Propagation Model of Laser Beams in Atmospheric Turbulence. *J. Opt. Soc. Am.* A2 (1985): 160–63.

18. Jakeman, E., and R.J.A. Tough. Generalised *K*-Distribution: A Statistical Model for Weak Scattering. *J. Opt. Soc. Am.* A4 (1987): 1764–72.

19. Barakat, R. Weak Scatter Generalisation of the *K*-Density Function with Application to Laser Scattering in Atmospheric Turbulence. *J. Opt. Soc. Am.* A3 (1986): 401–09.

20. Tough, R.J.A., D. Blacknell, and S. Quegan. A Statistical Description of Polarimetric and Interferometric Synthetic Aperture Radar Data. *Proc. Roy Soc.* A449 (1995): 567–89.

21. Tough, R.A.J. A Fokker-Planck Description of K-Distributed Noise. *J. Phys. A* 20 (1987): 551–67.

22. Field, T.R., and R.J.A. Tough. Stochastic Dynamics of the Scattering Amplitude Generating *K*-Distributed Noise. *J. Math. Phys.* 44 (2003): 5212–23.

23. Oliver, C.J. On the Simulation of Coherent Clutter Textures with Arbitrary Spectra. *Inverse problems* 3 (1987): 463–75.

24. Field, T.R., and R.J.A. Tough. Diffusion Processes in Electromagnetic Scattering Generating *K*-Distributed Noise. *Proc. Roy. Soc.* A 459 (2003): 2169–93.

25. Ward, K.D., R.J.A. Tough, and S. Watts. *Sea Clutter, the K-distribution and Radar Performance.* London: IEE, 2006.

13

Measurement and Detection

13.1 Introduction

In real experiments there are always a number of factors in the measuring process that modify the statistical properties of fluctuations in scattered waves so that they may appear to deviate from the expected behavior. In addition to the choice of scattering configuration, these include experiment duration, finite temporal and spatial resolution, and limited dynamic range.

Since all experiments last a finite time, it is not possible for the data to explore all values that the signal may take. The result of any measurement is then simply an *estimate* that will be subject to statistical errors and that may be *biased*. In other words, the results of a number of measurements will exhibit scatter about a value that may differ from the true value.

Limited detector resolution will change the statistical and correlation properties of the raw signal and, if sufficiently coarse, spatial and temporal integration may ultimately average out any fluctuations. However, it cannot be assumed that the fluctuations in an arbitrary signal will be reduced for *all* finite resolutions and indeed counter examples can be constructed where the degree of fluctuation initially increases as the resolution is coarsened. Finite dynamic range will also change the signal statistics in a way that is particular to the measuring instrumentation. This is usually a highly non-linear effect and may take the form of deletion or saturation in the presence of extreme fluctuations.

In addition to the effects pointed out above, noise will generally be introduced by the detection and signal processing hardware. This will often be additive and its properties can be measured in the absence of the signal. However, at optical frequencies the quantum nature of light will be exhibited in the detection process in the form of discrete photoelectric emissions; in this case, the postdetection signal is fundamentally digital and will be corrupted by photon noise.

In this chapter the implications of these experimental artifacts for the measured signal statistics will be explored for some of the models considered in earlier chapters, and their effect on measurement and detection will be discussed. These are important considerations because it is the fluctuations of the electrical signal at the output of the receiver rather than those of the raw signal that govern both the *performance* of detection systems designed to identify the presence of a specific signal, and the *accuracy* of measuring systems designed to determine a characteristic parameter of the signal.

13.2 Temporal Averaging of a Gamma-Lorentzian Process

As mentioned above, the effect of spatial or temporal integration of a signal is, *generally speaking*, to average out the fluctuations. In the case of a time-varying Gaussian process, the signal can be represented by the sum (Equation 1.23), namely

$$V(t) = \sum_{n=-\infty}^{\infty} v_n \exp(in\omega t) \quad where \quad \omega = \frac{2\pi}{T}$$

$$and \quad v_n = \frac{1}{T} \int_{-T/2}^{T/2} V(t) \exp(-in\omega t) dt$$

where the v_n are independent Gaussian variables. Integration of this quantity over a finite time interval or a memory function (the Fourier transform of a filter in frequency space) merely modifies the coefficients in the sum. This remains a weighted sum of Gaussian variables and is therefore still Gaussian but with a reduced variance and modified spectrum.

A similar argument can be applied to averaging in the spatial domain. However, when a nonlinear function of a Gaussian process is averaged, then the single interval statistics of the fluctuations are changed as well as the spectrum. Thus a number of exact results have been obtained for the effect of temporal integration on chi-square variables (Equation 3.11) constructed from a Gaussian process with Lorentzian spectrum. These results are special cases of those obtained for the effect of temporal integration on the gamma-Markov process governed by Equation 12.24 and Equation 12.27 and can be calculated from the generating function [1]

$$L(s, s', t; T) = \left\langle \exp\left[-sI(t+T) - s'E(t; T)\right] \right\rangle \tag{13.1}$$

Here the integrated variable E is defined in terms of the instantaneous gamma-Markov process *I* and the integration time *T* by

$$E(t;T) = \int_{t}^{t+T} I(t')dt' \tag{13.2}$$

It is not difficult to verify using the Laplace transform (12.7) of (12.24) that L satisfies the partial differential equation

$$\frac{\partial L}{\partial T} = -s\left[v + (\mu - \lambda + \lambda s)\frac{\partial}{\partial s}\right]L + s'\frac{\partial L}{\partial s} \tag{13.3}$$

This must be solved subject to the boundary condition $L(s, s', t; 0) = \langle \exp(-sI(t))\rangle$ that is given by Equation 12.9. The desired result is obtained by setting $s = 0$ in the solution of Equation 13.3 and it is readily shown that this has the form [1]

$$L(0,s',t;T) = \exp(\alpha\gamma)\Big/\left[\cosh y + \left(y/2\gamma + \gamma/2y\right)\sinh y\right]^{\alpha} \tag{13.4}$$

where $\alpha = v/\lambda$, $\gamma = (\mu - \lambda)T/2$, $y^2 = \gamma^2 + \lambda T^2 s'$. When $\alpha = n/2$, where n is a positive integer, this result coincides with that for the integrated statistics of chi-square variables constructed from the sum of the squares of n identical Gaussian-Lorentzian processes. The case $n = 1$ was first solved by Slepian [2] and the case $n = 2$ (that is relevant to thermal light and Gaussian speckle) by Bedard [3] using a method based on the properties of the underlying Gaussian variables that is only applicable to half integer values of α. It is not difficult to show from Equation 13.4 that the variance of the fluctuations decreases as the integration time increases, while the statistics of the fluctuations evidently change from chi-square.

13.3 Approximations for the Effect of Temporal and Spatial Integration

It is possible in principle to calculate the distribution corresponding to Equation 13.4 and its moments. The moments can be obtained by evaluating the derivatives of the generating function at $s' = 0$. Thus the second normalized moment may be expressed in the form

$$\frac{\langle E^2 \rangle}{\langle E \rangle^2} = 1 + \frac{1}{2\alpha\gamma^2}\left[2\gamma + \exp(-2\gamma) - 1\right] \tag{13.5}$$

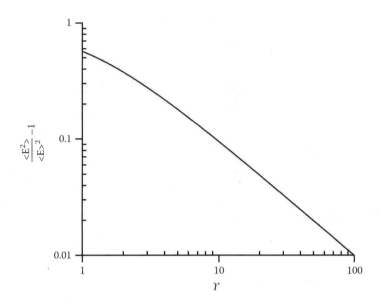

FIGURE 13.1
Normalized second moment of integrated intensity fluctuations for a gamma Lorentzian process (result of Equation 13.5).

This function is plotted in Figure 13.1 and shows a monotonic decrease in the degree of fluctuation with integration time, that is, the fluctuations are averaged out. When T is much larger than the fluctuation time $(\mu - \lambda)^{-1}$, (i.e., $\gamma \gg 1$) the integral (Equation 13.2) can be approximated by the sum of independent elements each of which is gamma distributed with index α. The generating function for such a sum is just equal to the product of the generating functions corresponding to each element. As in the case of the K-distributions considered in the previous chapter (Equation 12.1 and Equation 12.2), gamma variables are infinitely divisible. Thus it might be expected that the distribution of an integrated gamma process will also be a gamma variate with an index that is increased by the number of independent elements. This conjecture is born out by the form of Equation 13.5 which reduces to $1 + 1/N_c\alpha$ for the case of large γ, where $N_c \sim (\mu - \lambda)T$ is the number of fluctuation times in the integration time and may be interpreted as the number of independent elements of the signal within the integration time.

No analytic form for the inverse Laplace transform of Equation 13.4 has been published, but an approximate form for the distribution can be based on the above argument for the large γ limit using the exact result for the second moment (Equation 13.5). Thus assuming that the distribution of the integrated signal is approximately gamma distributed with index $\beta(T)$

$$P(E) = \frac{b^\beta E^{\beta-1}}{\Gamma(\beta)} \exp(-bE) \qquad (13.6)$$

$$\beta = \alpha\gamma^2 / \left[2\gamma + \exp(-2\gamma) - 1 \right] \tag{13.7}$$

The form of the distribution with this index will now be correct in the limit of small and large integration times and will have the correct second moment for *all* integration times. This type of approximation can be used to allow for spatial or temporal integration in the case of any process governed by an infinitely divisible distribution that is characterized by a single length/time scale. The time-dependent gamma-Markov process is one of the few cases for which a closed form solution for the generating function of the integrated statistics can be obtained, but the integrated second moment required for the approximation procedure can always be found directly from the auto-correlation function of the process. For example Equation 13.7 may be calculated from the identity

$$\left\langle E^2 \right\rangle = \int_0^T dt \int_0^T dt' \left\langle I(t)I(t') \right\rangle \tag{13.8}$$

In the case of an arbitrary response (memory) function or recursive filter M this can be written more generally in the form

$$\left\langle E^2 \right\rangle = \int_{-\infty}^{t} dt_1 \int_{-\infty}^{t} dt_2 M(t_1 - t)M(t_2 - t)\left\langle I(t_1)I(t_2) \right\rangle \tag{13.9}$$

The dependence on local time t vanishes in Equation 13.9 if the intensity is a stationary process. In frequency space the effect of filtering is accomplished by multiplying the spectrum of the process with that of the filter so that an alternative way of calculating the required result is to evaluate the total spectral energy (Equation 1.25). As already mentioned, the same approach can be used to approximate the effects of *spatial* integration on the statistics of infinitely divisible processes provided they are characterized by a single length scale.

13.4 Averaging Signals with More than One Scale

We have been careful in the deliberations above to restrict the approximation procedure to cases where the process is infinitely divisible and single scale. It is clear that if the process is not infinitely divisible then the distribution of the integrated signal may depart significantly from the distribution class of the original signal. Knowledge of the integrated second moment may then be insufficient to accurately characterize the evolution of the statistics.

The situation that arises in the case of a signal that is infinitely divisible but characterized by *more* than one scale may be investigated by considering the case of K-distributed noise. If the K-distributions arise as the gradient of a gamma process characterized by a single scale, or if the integration is carried out over an interval/area that is much larger than the largest scale present in the fluctuations, then the approach outlined above may be adopted with the help of Equation 12.2. On the other hand, it was shown in the previous chapter that K-distributed noise could be caused by the slow modulation of a fast-changing Gaussian speckle pattern. In this case the Gaussian (interference) contributions to the fluctuations can be attenuated by aperture averaging, low-pass filtering (i.e., temporal integration) or broadband illumination, (white light, frequency diversity) but the slow modulation of this contribution may remain. The *partially* smoothed signal will then fluctuate according to a gamma distribution as indicated in the discussion following Equation 12.18. In practical applications more sophisticated techniques have been developed to cope with these residual non-Gaussian fluctuations. For example, maritime radar systems are frequently limited by spurious returns from the sea surface that are effectively modulated speckle exhibiting K-distributed amplitude fluctuations. These systems often employ signal-processing algorithms that use some kind of adaptive threshold, based on an estimate of the K-distribution shape parameter, to eliminate the effect of the slow fluctuations. In cell-averaging constant false alarm rate (CFAR) detectors [4], the test range cell is compared with a weighted sum of returns from surrounding cells, assuming pulse-to-pulse frequency diversity is used to decorrelate the speckle contribution.

13.5 Enhancement of Fluctuations Caused by Filtering

In the foregoing section we have considered examples where the signal fluctuations are averaged out by temporal and spatial integration. However, in some situations filtering may actually *cause* fluctuations. Aperture averaging in heterodyne measurements, where the required information lies in the phase of the signal, is a case of particular practical importance. A tractable analog in the time domain is provided by the temporal averaging of a random phasor. A recursive filter operating on such a signal may be expressed in the form [5]

$$E(t) = \int_{-\infty}^{t} dt' M(t - t') \exp\left[i\varphi(t') \right] \tag{13.10}$$

As in Equation 13.9, M is a memory function that decays away into the past. The raw signal is the final exponential term under the integral sign so that as the memory function becomes very sharp the integration vanishes and the *amplitude* of E is constant.

Comparison with Equation 5.16 shows that Equation 13.10 is the time analog of a one-dimensional phase screen with aperture function M. Just as in the case of phase screen scattering, E develops amplitude fluctuations as the characteristic memory time increases. For example, if the phase in Equation 13.10 is a random function whose variance is much larger than unity, then the intensity of E changes from being constant when the memory function is very narrow to become a Gaussian process when the memory function spans many phase coherence times. In between these two limits, the character of the fluctuations is determined by the nature of the phase spectrum. For the case of a smoothly varying function, the variance of the fluctuations of $|E|^2$ will exceed that of a circular Gaussian process as found in Chapter 6. Figure 13.2 shows a numerical simulation in the case of a negative exponential memory function (Lorentzian filter) in Equation 13.10 assuming that the phase is a Gaussian random process with Gaussian spectrum. The peaks followed by small ripples can be interpreted as arising from the occurrence of stationary points in the phase function within the aperture/memory function leading to "diffraction-broadened caustics" in the filtered signal.

In practice, amplitude fluctuations will usually accompany the phase fluctuations appearing on the right-hand side of Equation 13.10, but these have only a modest effect on the outcome of the filtering process by comparison with the phase fluctuations. It is clear from this simple example that the effect of integration on non-Gaussian processes must be evaluated with considerable care.

13.6 The Effect of Finite Dynamic Range

Any real detection system will have a finite dynamic range. This may be deliberately designed to remove large excursions of the signal, for example in frequency demodulation where only the phase of the signal is of interest, and limiting may be used to remove amplitude fluctuations. In other cases it may be designed to encompass only the most frequently encountered excursions of the signal, for example if the required information is contained in the signal amplitude fluctuations. Probably the most common constraint on dynamic range is the limited number of bits used in analog-to-digital conversion. In all these cases there will be some amplitude limiting, either deliberate or inadvertent, and it is important to be aware of the likely consequences for the statistics of the measured signal fluctuations.

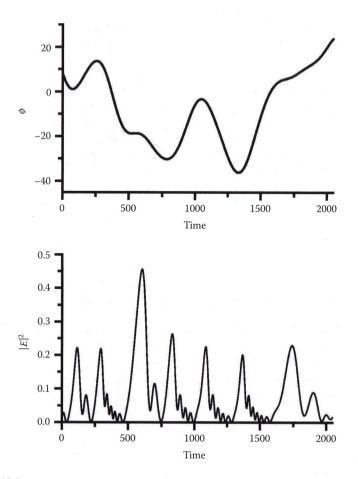

FIGURE 13.2

Lorentzian filtered time series: Gaussian random phasor with Gaussian spectrum (a) initial phase function, (b) intensity after filtering [5].

The simplest case to consider is that of Gaussian speckle where the intensity fluctuations are governed by a negative exponential distribution $b \exp(bI)$. If for simplicity it is assumed that the detector has constant sensitivity up to some maximum value X where there is a cutoff, then the truncated probability density must be normalized by the function $1 - \exp(-bX)$. The calculated values of the normalized moments can be expressed in terms of the incomplete gamma function

$$\frac{\langle I_L^n \rangle}{\langle I_L \rangle^n} = \frac{\left[1 - \exp(-bX)\right]^{n-1}}{\left[1 - \exp(-bX) - bX \exp(-bX)\right]^n} \gamma(n+1, bX) \qquad (13.11)$$

It is clear that the limited dynamic range leads to a significant reduction in the apparent degree of fluctuation.

A similar calculation can be carried out in the case of K-distributed fluctuations. The corresponding probability of finding the intensity I in the range $0 < I < X$ is given by the integral of definition (Equation 3.25) over this interval

$$P(0 < I < X) = 1 - \frac{2}{\Gamma(v)} \left(\frac{b\sqrt{X}}{2} \right)^v K_v(b\sqrt{X}) \tag{13.12}$$

Denoting the right-hand side of Equation 13.12 as $1/A$ we find [6]

$$\langle I^n \rangle = \frac{An!}{b^{2n}\Gamma(v)2^{v-1}} \left\{ 2^{2n+v-1}\Gamma(v+n) - \sum_{r=0}^{n} \frac{2^r(b\sqrt{X})^{v+2n-r}}{(n-r)!} K_{v+r}(b\sqrt{X}) \right\} \tag{13.13}$$

A plot of the second moment against X shows that truncation of the distribution again appears to cause a significant reduction in the degree of fluctuation (Figure 13.3). However, the relationship between the higher moments and the second remains relatively unaffected.

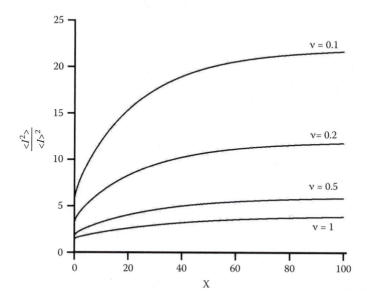

FIGURE 13.3
Effect of limited detector response on the normalized second moment of a K-distribution for the values of v shown and $b = 1$ (Equation 13.13).

It should be emphasized that the choice of limiting process used to calculate the results above will not be appropriate in every case. Although a sharp cutoff may often be usefully imposed to avoid unacceptable or unknown nonlinear response of the detection hardware in the presence of large signal excursions, it removes the possibility of correcting for errors produced by such nonlinearity.

13.7 The Effect of Finite Measurement Time

After making allowance for instrumental effects, the most important consideration in analyzing the performance of detection and measuring systems is the effect of a finite measurement time. Generic examples will serve to illustrate the method of approach that can be used to evaluate this effect for the case of a measuring system. Suppose, for example, that a measurement of the autocorrelation function (or equivalently, the spectrum) of a stationary, ergodic process $I(t)$ is required. An estimate of this quantity is constructed experimentally by adding sample products of the process separated by the selected delay time

$$\hat{G}^{(2)}(\tau) = \frac{1}{N} \sum_{n=1}^{N} I(t_n + \tau) I(t_n) \tag{13.14}$$

The times $\{t_n\}$ may be chosen at random, or may be chosen to have a regular period p. The estimator (Equation 13.14) is said to be *unbiased* because

$$\left\langle \hat{G}^{(2)}(\tau) \right\rangle = \frac{1}{N} \sum_{n=1}^{N} \left\langle I(t_n + \tau) I(t_n) \right\rangle = \left\langle I(0) I(\tau) \right\rangle \tag{13.15}$$

In other words, the average of the estimator is equal to the required correlation function. If the experiment time is much larger than the correlation time of the process and N is large so that the sum of Equation 13.14 contains many independent samples, then the estimator will be approximately Gaussian-distributed according to the central limit theorem. The variance of the estimator should then provide a good guide to the accuracy of the measure. It is not difficult to show that

$$\operatorname{var}\hat{G}^{(2)}(\tau) = \frac{1}{N}\operatorname{var}\left[I(0)I(\tau)\right]+$$

$$+\frac{1}{N^2}\sum_{n\neq m=0}^{N}\left[\left\langle I(t_n+\tau)I(t_n)I(t_m+\tau)I(t_m)\right\rangle-\left(G^{(2)}(\tau)\right)^2\right] \tag{13.16}$$

Taking advantage of stationarity and the symmetry of the terms in the sum, and writing $n = kp$ where p is the sampling period, this may be reduced to the form [7]

$$\operatorname{var}\hat{G}^{(2)}(\tau) = \frac{1}{N}\operatorname{var}\left[I(0)I(\tau)\right]+$$

$$+\frac{1}{N}\sum_{k=1}^{N-1}\left(1-\frac{k}{N}\right)\left[\left\langle I(kp+\tau)I(kp)I(\tau)I(0)\right\rangle-\left(G^{(2)}\right)^2(\tau)\right] \tag{13.17}$$

The sum arises as a consequence of correlation between the samples. If the sampling period is much longer than the correlation time, then its contribution will be small because $\langle I(kp + \tau)I(kp)I(\tau)I(0)\rangle \approx (G^{(2)})^2$ so that

$$\operatorname{var}\hat{G}^{(2)}(\tau) \approx \frac{1}{N}\left(\left\langle I^2(0)I^2(\tau)\right\rangle-\left\langle I(0)I(\tau)\right\rangle^2\right) \tag{13.18}$$

This can be evaluated rather easily if I is the intensity of a zero-mean complex Gaussian process, for example (see Equation 2.21 and Equation 2.25)

$$\operatorname{var}\hat{G}^{(2)}(\tau) = \frac{3\left(G^{(2)}(\tau)\right)^2+8G^{(2)}(\tau)-8}{N} = \frac{\langle I\rangle^4}{N}\left[3+14\left|g^{(1)}(\tau)\right|^2+3\left|g^{(1)}(\tau)\right|^4\right] \tag{13.19}$$

It is not difficult to check that the expression in square brackets reduces to the correct limits of 20 and 3 when τ is zero and infinity, respectively. The most relevant formulation of this result is the ratio of the standard deviation of the measurement to its mean value. This shows that the accuracy is inversely proportional to the square root of the number of samples

$$\frac{\sqrt{\operatorname{var}\hat{G}^{(2)}(\tau)}}{G^{(2)}(\tau)} = \frac{1}{\sqrt{N}}\frac{\sqrt{3+14\left|g^{(1)}(\tau)\right|^2+3\left|g^{(1)}(\tau)\right|^4}}{1+\left|g^{(1)}(\tau)\right|^2} \tag{13.20}$$

A plot of Equation 13.20 for the case of a Lorentzian spectrum, $g^{(1)}(\tau) = \exp(-\Gamma\tau)$, shows little variation with delay time, with the relative error decreasing monotonically from $\sqrt{5/N}$ to $\sqrt{3/N}$ over the range $[0, \infty]$.

It is desirable in some measurements to adopt a technique for removing drift or long-term variations in the experimental conditions. In the correlation measurement (Equation 13.14), for example, it may be advantageous to directly construct the normalized function

$$\hat{g}^{(2)}(\tau) = \frac{\dfrac{1}{N}\displaystyle\sum_{n=1}^{N} I(t_n + \tau)I(t_n)}{\left(\dfrac{1}{N}\displaystyle\sum_{n=1}^{N} I(t_n)\right)^2} \tag{13.21}$$

Here the denominator is equal to the square of an unbiased estimate of the mean that can be written in the form

$$\hat{I} = \frac{1}{N}\sum_{n=1}^{N} I(t_n) = \langle I \rangle + \frac{1}{N}\sum_{n=1}^{N}\left[I(t_n) - \langle I \rangle\right] \tag{13.22}$$

It is clear that the average value of Equation 13.21 is not equal to the desired correlation function; this kind of measurement is a *biased* estimate. If N is large, the sum on the right of Equation 13.22 is much smaller than $\langle I \rangle$ and the denominator may be expanded so that

$$\hat{g}^{(2)}(\tau) \approx \frac{1}{N\langle I\rangle^2}\sum_{n=1}^{N} I(t_n + \tau)I(t_n)\left[1 - \frac{2}{N\langle I\rangle}\sum_{m=1}^{N}\left[I(t_n) - \langle I\rangle\right] + \cdots\right] \tag{13.23}$$

Averaging this equation shows that the bias correction is inversely proportional to the number of samples. Thus the bias correction is much smaller than the standard deviation of the estimate and can be ignored provided that N is large. Calculation of the mean and variance of the estimate (Equation 13.21) for Gaussian-Lorentzian intensity fluctuations (gamma-Markov process with unit index) [8] shows that this is indeed the case and that, moreover, the relative standard deviation is actually *reduced* by comparison with the unbiased result (Equation 13.20). This is due to some cancelling of the fluctuations in the numerator and denominator of Equation 13.21.

These arguments work well for "well-behaved" fluctuations, giving relative standard deviations of order 1% and bias of order 0.01% for 10,000 samples. However, in the case of more wildly fluctuating signals, even unbiased estimates can give misleading results unless the number of samples is exceedingly large [9]. The following simple argument shows why an unbiased estimate of a statistical moment can give inaccurate results. Suppose the data consists of N uncorrelated samples x_i from a probability distribution $p(x)$, then an unbiased estimate of the rth moment $M^{(r)}$ is given by

$$\hat{M}^{(r)} = \frac{1}{N} \sum_{i=1}^{N} x_i^r \tag{13.24}$$

The variance of this average is just $1/N$ times the variance of each term (see derivation of Equation 13.18)

$$\operatorname{var} \hat{M}^{(r)} = \frac{1}{N} \left[\left\langle x^{2r} \right\rangle - \left\langle x^r \right\rangle^2 \right]$$

$$= \frac{1}{N} \left[M^{(2r)} - \left(M^{(r)} \right)^2 \right] \tag{13.25}$$

Clearly for a reasonable estimator of a moment to be obtained, its variance must be small compared with the square of its mean. Therefore, it is required that

$$\frac{1}{N} \left[M^{(2r)} - \left(M^{(r)} \right)^2 \right] \ll \left(M^{(r)} \right)^2 \tag{13.26}$$

or

$$N \gg \frac{M^{(2r)}}{\left(M^{(r)} \right)^2} - 1 \tag{13.27}$$

Unfortunately, the sample size given by the right-hand side of this relation can be very large indeed. For the fifth moment of the lognormal distribution with moments given by Equation 3.29 and with a second moment of seven, for example, it is $7^{25} = 1.3 \times 10^{21}$! Even for the relatively well-behaved K-distribution (Equation 3.25), the right-hand side of Equation 13.27 is 2.5×10^5 for the fifth moment when the second moment is seven. This may be contrasted with the negative exponential distribution characterizing a conventional Gaussian speckle pattern for which Equation 13.27 requires only that $N \gg 25$.

When the sample size is smaller than that prescribed above, then the standard deviation of the moment estimate is larger than the mean, and since the moment estimate must be positive, this implies a highly skewed distribution with a median smaller that the mean and with a long tail. Thus, estimates of the moments will be biased towards low values despite the fact that the average value of Equation 13.24 is the true mean.

For simplicity, correlation between samples in the calculations above has been consistently neglected. In practice, maximum amount of information can be extracted from the signal when the samples are contiguous and this will usually imply that the samples are correlated. The sums on the right-hand side of Equation 13.16 and Equation 13.17 cannot then be neglected. In the case of a gamma-Markov process, these terms can be evaluated exactly to give rather lengthy formulae that will not be reproduced here. However, they show that the reduction in the standard deviation of the estimate obtained by utilizing the additional correlated samples is rather modest [8]. However, the conclusions above are not valid in the case of a power-law spectrum with no outer scale because the correlation between samples may decrease slowly over many decades, leading to a very slow improvement in accuracy with experiment time for some ranges of the power-law index [10].

13.8 Noise in Frequency Demodulation

In previous sections, measurement errors have arisen due to some kind of averaging, limited dynamic range, or finite measurement time. However, in some situations it is possible to improve the accuracy of measurements by utilizing information in the signal that would normally be discarded. This is illustrated by the measurement of a simple frequency shift f in a carrier that is modulated by a complex Gaussian process

$$S(t) = A(t)\exp\left\{i\left[\phi(t)+ft\right]\right\} \tag{13.28}$$

This situation would be encountered, for example, in the heterodyne measurement of light that had been reflected from a vibrating rough surface. The response of an ideal frequency demodulator may be written as the phase derivative

$$\dot{\theta} = \dot{\phi}+f \tag{13.29}$$

This equation is in the conventional "signal plus noise" format where the signal is the required frequency shift and the noise is the phase derivative of a complex Gaussian process. However, it was shown in Chapter 2 that the fluctuations of the phase derivative of a complex Gaussian process are governed by a student's t distribution (Equation 2.38) whose variance is infinite. Thus any frequency measurement based on Equation 13.29 will be susceptible to very large fluctuations. This problem arises owing to the large phase jumps that can occur when the amplitude A in Equation 13.28 is zero and may be overcome by weighting the phase derivative with the amplitude in constructing the estimator [11]

$$\hat{f} = \frac{\sum_{n=1}^{N} A_n^{\nu} \dot{\theta}_n}{\sum_{n=1}^{N} A_n^{\nu}} = f + \frac{\sum_{n=1}^{N} A_n^{\nu} \dot{\phi}_n}{\sum_{n=1}^{N} A_n^{\nu}} \qquad (13.30)$$

Evidently $\langle \hat{f} \rangle = f$ so that $\hat{f}$ is an unbiased estimate of the frequency offset. Taking advantage of the statistical independence of the samples and using the result that $A\dot{\phi}$ is independent from any function of A (see Section 2.8) obtains

$$\text{var}\,\hat{f} = \left\langle \frac{\sum_{n=1}^{N} A_n^{2(\nu-1)}}{\left(\sum_{n=1}^{N} A_n^{\nu}\right)^2} \right\rangle \left\langle \left(A\dot{\phi}\right)^2 \right\rangle \qquad (13.31)$$

It is not difficult to demonstrate analytically that this quantity has a minimum when $\nu = 2$ so that the measurement error is minimized by using the intensity-weighted phase derivative in the estimator (Equation 13.30). The right-hand side of Equation 13.31 is then inversely proportional to the sum of N exponentially distributed independent variables and can be evaluated using the appropriate chi-square distribution and results in Section 2.8 to give var $\hat{f} = [2(N-1)\tau_c^2]^{-1}$.

13.9 Detection

Signal detection in the presence of noise is a classic radar problem that figures widely in the literature [12]. Processing of the postdetection signal, which may be contaminated with multiplicative as well as additive noise, is always based on exploitation of differences between the signal and the noise. These may relate, for example, to spatial and temporal fluctuation scales, to the statistics of the fluctuations, or to polarization properties. The detailed analysis of performance is particular to the scenario of interest and here only an outline will be presented of the simplest generic problem, namely, signal detection in the presence of additive noise, to demonstrate the application of fluctuation models in this context. It will be assumed that only the intensity of the wave can be measured, that is, direct detection so that

$$I = I_N \qquad \textit{noise only}$$

$$I = I_S + I_N \quad \textit{signal + noise}$$

(13.32)

In a conventional analysis of the detection problem a decision on the presence or otherwise of the signal is based on a threshold criterion

$$I > k \Rightarrow \textit{signal present} \tag{13.33}$$

The threshold is set on the basis of some *a priori* knowledge of the noise level. Suppose that the time available for measurement is T_{exp}. A single number corresponding to the total intensity received during this period may be obtained, or the time may be split into several measurements. The choice depends on a number of factors and, as already mentioned, the object is to exploit differences in the characteristics of the signal and noise. However, if the fluctuation times of both signal and noise are greater than the experiment time and there is no other means by which they may be distinguished, other than by their respective intensities, then there is little to be gained from sophisticated processing techniques. In this case the performance of the detection algorithm Equation 13.32 is governed by the first-order probability densities of the signal and noise and is usually characterized by three measures that depend on the threshold: the detection probability p_d, the missed signal probability p_m, and the false alarm probability p_f. These are given in terms of the statistical properties of the signal and noise by (Figure 13.4)

$$p_d = prob(I_{S+N}) > k = \int_k^\infty P_{S+N}(I)dI$$

$$p_m = 1 - p_d \tag{13.34}$$

$$p_f = prob(I_N) > k) = \int_k^\infty P_N(I)dI$$

and can be evaluated to give a prediction of expected performance provided that realistic models for the fluctuations of both signal and noise are available. The result will depend on the integrated intensity that is proportional

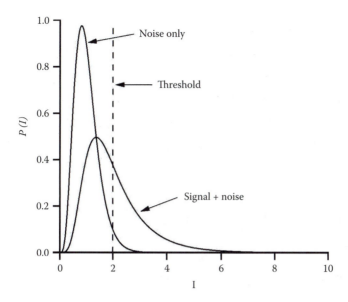

FIGURE 13.4
Threshold detection.

to the experiment time, as well as on any other parameters of the models. As an example, consider a situation in which the signal intensity fluctuates according to a negative exponential distribution while the noise is governed by a gamma distribution

$$P_N(I_N) = \frac{1}{\Gamma(\alpha)} \left(\frac{\alpha I_N}{\langle I_N \rangle} \right)^{\alpha-1} \frac{\alpha}{\langle I_N \rangle} \exp\left(-\frac{\alpha I_N}{\langle I_N \rangle} \right) \qquad (13.35)$$

If it is assumed that the signal and noise intensities are statistically independent and additive, then the generating function for the distribution of the total intensity when the signal is present is given by the product of the generating functions for the distributions corresponding to the signal and noise intensity alone

$$Q_{S+N}(s) = \langle \exp(-sI) \rangle = \left(1 + \langle I_S \rangle s \right)^{-1} \left(1 + \langle I_N \rangle s / \alpha \right)^{-\alpha} \qquad (13.36)$$

The corresponding distribution contains an incomplete gamma function [13]

$$P_{S+N}(I) = \frac{1}{\langle I_S \rangle \Gamma(\alpha)} \frac{\exp\left(-I/\langle I_S \rangle\right)}{\left(1 - \langle I_N \rangle / \alpha \langle I_S \rangle\right)^{\alpha}} \gamma\left[\alpha, \left(1 - \langle I_N \rangle / \alpha \langle I_S \rangle\right)\alpha I / \langle I_N \rangle\right]$$

$$= \frac{\exp\left(-I/\langle I_S \rangle\right)}{\langle I_S \rangle \Gamma(\alpha)} \left(\frac{\alpha I}{\langle I_N \rangle}\right)^{\alpha} \sum_{n=0}^{\infty} \frac{1}{n!(\alpha+n)} \left[\left(\frac{\alpha}{\langle I_N \rangle} - \frac{1}{\langle I_S \rangle}\right) I\right]^{n}$$

$$(13.37)$$

Thus $(\Gamma(\alpha, x) = \Gamma(\alpha) - \gamma(\alpha, x)$ [13])

$$p_d = \frac{1}{\Gamma(\alpha)} \Gamma\left(\alpha, \frac{\alpha k}{\langle I_N \rangle}\right) + \frac{\exp\left(-k/\langle I_S \rangle\right)}{\Gamma(\alpha)\left(1 - \langle I_N \rangle / \alpha \langle I_S \rangle\right)^{\alpha}} \gamma\left(\alpha, \left(1 - \frac{\langle I_N \rangle}{\alpha \langle I_S \rangle}\right)\frac{\alpha k}{\langle I_N \rangle}\right)$$

$$p_f = \frac{1}{\Gamma(\alpha)} \Gamma\left(\alpha, \alpha k / \langle I_N \rangle\right)$$

$$p_m = 1 - p_d$$

$$(13.38)$$

Note that as $\alpha \to \infty$, $p_d \to \exp[(\langle I_N \rangle - k)/\langle I_S \rangle]$, while $p_f \to 0$. In this limit the fluctuations in the noise vanish, and so there is no chance of a false alarm. Moreover, the probability of detection is just the area under the tail of the signal negative exponential distribution that lies above the threshold, which must obviously be set above the constant noise level.

Results of Equation 13.38 may be expressed in terms of three parameters: the signal-to-noise ratio, $S = \langle I_S \rangle / \langle I_N \rangle$, the threshold-to-noise ratio $K = k/\langle I_N \rangle$, and the relative variance of the noise α^{-1}. Figure 13.5 shows how the false alarm and missed signal probabilities vary with threshold level, K, for typical values of α and S. Notice that the total error, $p_e = p_f + p_m$, has a minimum for a particular threshold setting.

For an integration time that is larger than the correlation time of a signal or noise process, the degree of fluctuation will generally be reduced. The calculation above can be adapted to model this situation by adopting a gamma distribution for the signal as well as the noise, and choosing values of the -parameters of the signal and noise models to be appropriate for the degree of averaging. In this way a dependence on pulse transmission duration can be incorporated in the calculation. A similar approach can be adopted in the evaluation of the effect of aperture averaging and illuminated area on performance.

Suppose now that N independent measurements are made; the probability of detection, based on finding R of these that exceed the threshold, is given in terms of the binomial coefficients by

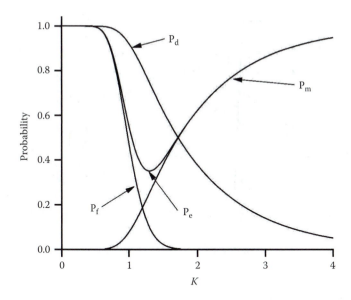

FIGURE 13.5
Dependence of false alarm and missed signal probabilities on threshold level: negative exponentially distributed signal in gamma distributed noise with $\alpha = 20$, $S = 1$.

$$P_d = 1 - \sum_{q=0}^{R-1} P_q = 1 - \sum_{q=0}^{R-1} {}^N C_q p_d^q (1 - p_d)^{N-q} \qquad (13.39)$$

A similar formula can be written down for the probability of a false alarm in terms of p_f. The simplest case is $R = 1$, that is, detection is presumed on finding the intensity of *one* sample to be greater than the threshold. Then

$$P_d = 1 - (1 - p_d)^N$$
$$P_m = (1 - p_d)^N \qquad (13.40)$$
$$P_f = 1 - (1 - p_f)^N$$

These quantities are plotted in Figure 13.6 for the case $N = 100$. As expected, there is a significant improvement in performance as measured by the broad region of low error probability. However, whether the time available for detection should be expended in a single long pulse or divided into a large number of samples depends critically on the characteristics of the noise and the performance criteria that have to be met.

Performance of detection and measuring systems will also depend on the choice of scattering geometry. For example, it was shown in Chapter 10 that the statistics of the return after double passage through a random medium

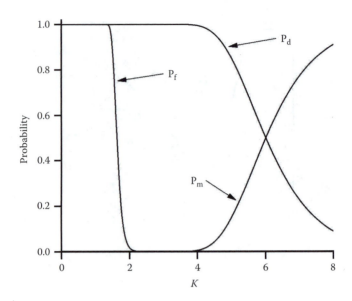

FIGURE 13.6
Improvement in performance achieved by using 100 independent samples.

may be different according to whether the radiation traverses the same or a different path after reflection from the target. The related detection problem has been analyzed for cases where, at the first pass, the target is illuminated by a signal that is either Gaussian speckle or K-distributed [14]. It is found that for low error probabilities and high signal-to-noise ratios, bistatic operation, in which scattering in the forward and reverse paths is uncorrelated, results in an improvement in detection performance that increases as the fluctuations of the signal increase. This is because signal fading or "drop out" is a more important consideration than signal intensity in this limit, and fading is less in bistatic configurations. However, for low signal-to-noise ratios monostatic operation, where there is correlation between the scattering on forward and reverse paths, may significantly reduce the probability of error because the signal intensity is now more critical and is greater in this configuration. The crossover is relatively model independent but occurs at signal-to-noise ratios that increase as the signal fluctuations increase (Figure 13.7).

13.10 Quantum Limited Measurements

In the measurement of light intensity, detectors are generally based on the process of photoelectric emission in which absorption of the incident photons causes electrons to be ejected. This means that at the output of the detector

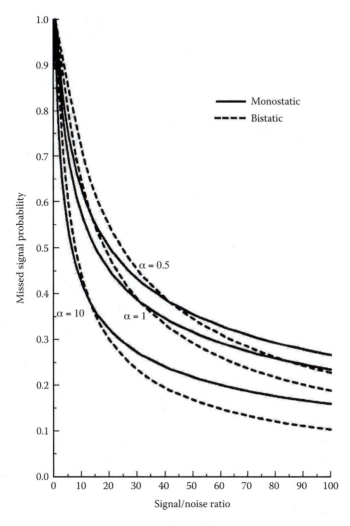

FIGURE 13.7
Comparison of monostatic and bistatic detection performance in double passage geometries; missed signal probability vs. signal-to-noise ratio with a false alarm ratio of 10^{-6}, K-distributed signal at the target ($v = 0.5$) in gamma distributed noise with shape parameters as shown [14].

the signal is fundamentally discrete in nature. In practice, the output of a photomultiplier or avalanche photodiode in Geiger mode, for example, takes the form of a set of pulses of varying height that are standardized by an amplifier and discriminator with a small percentage loss [15]. The signal processor is thus presented with a train of identical events or "photon" arrivals. Actually these are electron emissions and occur at a lower rate than the incident photon arrivals due to the inefficiency of the detector that fails to respond to a subset of the photons. A full quantum mechanical treatment of detection is essential for the correct interpretation of experiments using so-called nonclassical and "squeezed" light [16] that may be used, for example, in research on cryptography and quantum computers. Although this topic is growing in importance, for the overwhelming majority of light scattering applications where the principle source of radiation is the laser, the semiclassical theory of detection is a perfectly adequate tool for the analysis of system performance. This theory represents photoelectric emission as a doubly stochastic Poisson process in response to a classical Maxwell field falling on the detector surface. The probability of registering n counts in the time interval T is then given by [17]

$$p(n;T) = \frac{1}{n!} \int_0^\infty dE (\eta E)^n \exp(-\eta E) P(E) \tag{13.41}$$

Here, E is the integrated intensity (Equation 13.3) of the incident optical field, and η is the quantum efficiency of the detector. It is not difficult to show that

$$\langle (1-s)^n \rangle = \langle \exp(-s\eta E) \rangle \tag{13.42}$$

Thus the generating function for the discrete process is the same as the generating function for the distribution of the integrated incident intensity allowing for detector quantum efficiency. An important implication of this result is that the factorial moments of the photon fluctuations are proportional to the moments of the intensity fluctuations

$$N^{[r]} = \langle n(n-1)(n-2)\cdots(n-r+1) \rangle = \eta^r \langle E^r \rangle \tag{13.43}$$

The photon-counting distributions associated with the fluctuation models that have been encountered earlier can be derived directly from Equation 13.41. For the case of a laser with $P(E) = \delta(E - E_0)$, a Poisson distribution is obtained (Figure 13.8)

$$p(n;T) = \frac{\bar{n}^n}{n!} \exp(-\bar{n}); \quad n^{[r]} = 1; \quad \bar{n} = \eta E_0 \tag{13.44}$$

Here $n^{[r]}$ is the rth *normalized* factorial moment obtained by dividing Equation 13.43 by $\bar{n}^r$, where $\bar{n}$ is the mean number of counts in the sample time T. For the case of thermal light (Gaussian speckle) $P(E) = \exp(-E/\langle E \rangle)/\langle E \rangle$ assuming that there is no spatial or temporal integration and Equation 13.37 predicts a geometric distribution (Figure 13.8)

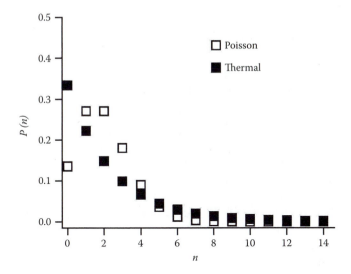

FIGURE 13.8
Distributions of photon counts with a mean count rate of two expected for (a) coherent light
(Equation 13.44) and (b) thermal light (Equation 13.45).

$$p(n;T) = \frac{\bar{n}^n}{(1+\bar{n})^{n+1}}; \quad n^{[r]} = n!; \quad \bar{n} = \eta\langle E \rangle \tag{13.45}$$

Note that in this result and the formulae below $\langle E \rangle = \langle I \rangle T$ according to
Equation 13.3 where I is the instantaneous intensity. For a gamma variate
that may characterize a space or time averaged speckle pattern, on the other
hand, the negative binomial distributions (Equation 12.10) are obtained

$$p(n;T) = \frac{\bar{n}^n}{\left(1+\dfrac{\bar{n}}{\alpha}\right)^{n+\alpha}}; \quad n^{[r]} = \frac{\Gamma(n+\alpha)}{\alpha^n\Gamma(\alpha)}; \quad \bar{n} = \eta\langle E \rangle \tag{13.46}$$

In optical homodyne experiments the received intensity is governed by the
Rice distribution (Equation 3.5). The photon-counting distribution and its
factorial moments can be expressed in terms of Laguerre polynomials for
this model

$$p(n;T) = \frac{\bar{n}_g^n}{(1+\bar{n}_g)^{n+1}} \exp\left\{-\frac{\bar{n}_c}{(1+\bar{n}_g)}\right\} L_n\left\{-\frac{\bar{n}_c}{\bar{n}_g(1+\bar{n}_g)}\right\}$$

(13.47)

$$N^{[r]} = r!\bar{n}_g^r L_r\left(-\frac{\bar{n}_c}{\bar{n}_g}\right); \quad \bar{n}_g = \eta\langle E_g\rangle; \quad \bar{n}_c = \eta\langle E_c\rangle$$

Some results have also been obtained for the case of heterodyning where a frequency shift is introduced between the scattered light and the local oscillator [18].

The photon-counting distribution corresponding to K-distributed intensity fluctuations can also be expressed in terms of special functions. Substituting the definition (Equation 3.24) into Equation 13.41 obtains a Whittaker function distribution [10]

$$p(n;T) = n!\bar{n}^{-\alpha/2} \frac{\Gamma(n+\alpha)}{\Gamma(\alpha)} \exp\left(\frac{\alpha}{2\bar{n}}\right) W_{-(n+\alpha/2),(\alpha-1)/2}\left(\frac{\alpha}{\bar{n}}\right)$$

(13.48)

$$N^{[r]} = \bar{n}^r r! \Gamma(r+\alpha); \quad \bar{n} = \eta\langle E\rangle = \eta T\alpha/b$$

This distribution has been invoked to account for data taken in photon-counting measurements of optical scintillation generated by thermal plumes and propagation through the turbulent atmosphere [19].

Equation 13.41 can be generalized to define the relationship between the joint intensity and joint photon-counting statistics

$$p(n,n';T) = \int_0^\infty dE \int_0^\infty dE'(\eta E)^n (\eta E')^{n'} \exp\left[-\eta(E+E')\right] P(E,E')$$ (13.49)

This quantity is rarely measured directly in experiments, but the corresponding correlation functions have figured prominently in the literature as a consequence of their importance in applications of light scattering techniques [15, 20]. Fortunately, it can be shown from Equation 13.49 that these are simply related to the corresponding intensity correlation functions. For example

$$\langle nn'\rangle = \eta^2 \langle EE'\rangle$$

(13.50)

This means that the calculation of experimental accuracy described above can be carried out for photon-counting measurements with little additional complication. However, care must be taken when products of intensities measured at the same time are invloved. Thus, for example

$$\langle n^2 n' \rangle = \langle n(n-1)n' \rangle + \langle nn' \rangle$$

$$= \eta^3 \langle E^2 E' \rangle + \eta^2 \langle E' \rangle \tag{13.51}$$

This can then be evaluated from the chosen intensity fluctuation model. Early experiments confirmed the accuracy of factorial moment measurements predicted by this approach [21] and a rigorous analysis of the accuracy of optical line-width measurements by photon correlation spectroscopy was subsequently carried out [8].

References

1. Jakeman, E. Statistics of Gamma Lorentzian Intensity Fluctuations. *Optica Acta* 27 (1980): 735–881.
2. Slepian, D. Fluctuations of Random Noise Power. *Bell Syst. Tech. J.* (January 1958): 163–84.
3. Bedard, G. Photon Counting Statistics of Gaussian Light. *Phys. Rev.* 151 (1966): 1038–39.
4. Watts, S., C.J. Baker, and K.D. Ward. Maritime Surveillance Radar Part 2: Detection Performance Prediction in Sea Clutter. *IEE Proceedings* 137 (1990): 63–72.
5. Jakeman, E., and K.D. Ridley. Signal Processing Analogue of Phase Screen Scattering. *J. Opt. Soc. Am.* A15 (1998): 1149–59.
6. Jakeman. *K*-Distributed Noise. *J. Opt. A: Pure Appl. Opt.* 1 (1999): 784–89.
7. Davenport, W.B., and W.L. Root. *An Introduction to the Theory of Random Signals and Noise.* New York: McGraw-Hill, 1958.
8. Jakeman, E., E.R. Pike, and S. Swain. Statistical Accuracy in the Digital Autocorrelation of Photon Counting Fluctuations. *J. Phys. A: Math. Gen.* 4 (1971): 517–34.
9. Tapster, P.R., A.R. Weeks, P.N. Pusey, and E. Jakeman. Analysis of Probability Density Functions for Laser Scintillations in a Turbulent Atmosphere. *J. Opt. Soc. Am.* A6 (1989): 782–85.
10. Jakeman, E., and D.L. Jordan. Statistical Accuracy of Measurements on Gaussian Random Fractals. *J. Phys. D: Appl. Phys.* 23 (1990): 397–405.
11. Jakeman, E., S.M. Watson, and K.D. Ridley. Intensity-Weighted Phase Derivative Statistics. *J. Opt. Soc. Am.* A18 (2001): 2121–31.
12. Skolnik, M.I. *Introduction to Radar Systems.* New York: McGraw-Hill,1962).
13. Abramowitz, M., and I.A. Stegun. *Handbook of Mathematical Functions.* New York: Dover, 1971.
14. Jakeman, E., J.P. Frank, and G.J. Balmer. The Effect of Enhanced Backscattering on Target Detection. AGARD Conference Proceedings 542 (1993): 32/1–7.
15. Oliver, C.J. Correlation Techniques. In H.Z. Cummins and E.R. Pike, eds. *Photon Correlation and Light Beating Spectroscopy.* New York: Plenum Press, 1974.
16. Loudon, R., and P.L. Knight. Squeezed Light. *J. Mod. Opt.* 34 (1987): 709–59.
17. Mandel, L. Fluctuations in Photon Beams: The Distribution of the Photo-electrons. *Proc. Phys. Soc.* 74 (1959): 233–43.

18. Jakeman, E., and E.R. Pike. Statistics of Heterodyne Detection of Gaussian Light. *J. Phys. A (Gen Phys)* 2 (1969): 115–25.
19. Parry, G., and P.N. Pusey. *K*-distributions in Atmospheric Propagation of Laser Light. *J. Opt. Soc. Am.* 69 (1979): 796–98.
20. Cummins, H.Z., and E.R. Pike, eds. *Photon Correlation Spectroscopy and Velocimetry*. New York: Plenum Press, 1977.
21. Jakeman, E., C.J. Oliver, and E.R. Pike. Measurement of the Factorization Properties of Higher Order Optical Correlation Functions. *J. Phys. A* 1 (1968): 497–99.

14

Numerical Techniques

14.1 Introduction

Numerical techniques play an important role in investigations of the fluctuations of scattered waves. This chapter covers two aspects: generation of random processes, and simulation of wave propagation. The former allows one to produce a set of numbers which conform to a particular statistical model, a *realization* of the random process under consideration, and the latter to investigate the propagation of radiation after it has interacted with a scattering system. For example, one can generate a set of numbers corresponding to the phase shifts produced at different positions on a random phase screen and then simulate the propagation of electromagnetic radiation through the screen to investigate the development of intensity fluctuations. This approach complements the analytical techniques described in the rest of this book. Analytical techniques usually calculate averages or probability distributions. With numerical techniques one can observe single realizations, which will often give useful physical insight into the scattering process. Averages, and other statistical quantities, can also be calculated numerically by repeatedly generating realizations. These can be used to check the results of analytical calculations and to investigate regimes that are analytically intractable (this also works in the reverse direction; in fact, the first thing one should always try to do with a new numerical algorithm is to reproduce a known analytical result). There are, of course, limits to what can be achieved numerically, and numerical techniques will never replace analysis.

14.2 The Transformation of Random Numbers

Many software packages have built-in random number generators (more correctly described as pseudorandom number generators). These usually allow the user to generate numbers which are uniformly distributed over some interval such as 0 to 1. For the purposes of this chapter it is assumed that such a generator exists and that the numbers it produces are sufficiently random. The interested reader is directed to Reference [1] for detailed discussion of these matters.

Random numbers conforming to other probability distributions can be obtained by applying a transforming function to uniformly distributed numbers. Consider the general case in which a function f is applied to a random number x to give a new number y. It was shown in Chapter 1 (Equation 1.11) that the probability densities of the original and transformed variables are related by

$$P_y(y)\left|\frac{df}{dx}\right| = P_x(x)$$ (14.1)

Here, subscripts have been used to distinguish the x and y probability densities. The transformation of numbers of pdf $P_x(x)$ to numbers of pdf $P_y(y)$ requires the solution of the differential Equation 14.1 to find f. Integrating both sides gives an equation in terms of the *cumulative* distribution functions, which are defined by Equation 1.3

$$F_y(y) = F_x(x)$$ (14.2)

So the required transforming function can be expressed as

$$y = f(x) = F_y^{-1}(F_x(x))$$ (14.3)

F_y^{-1} being the inverse of the cumulative distribution function of y.

When the variable x is uniformly distributed between 0 and 1, $F_x(x) = x$ and the required transformation is

$$y = f(x) = F_y^{-1}(x)$$ (14.4)

14.3 Gaussian Random Numbers

Gaussian distributed random numbers may be generated from uniformly distributed ones via Equation 14.4 using the inverse error function, but in practice it is more computationally efficient to use a scheme which does not involve special functions. A useful algorithm is based on the polar method [1]. This starts with two random numbers, x_1 and x_2, which are uniformly distributed between 0 and 1, and transforms them to give two numbers, y_1 and y_2, which are the Cartesian components of a two-dimensional, zero-mean, circular Gaussian distribution (see Equation 2.2). The transformation uses polar coordinates A and ϕ ; x_1 is transformed to give a uniformly distributed phase angle ϕ (in radians)

$$\phi = 2\pi x_1 \tag{14.5}$$

and x_2 is transformed into a negative exponentially distributed random variable, the square root of which gives the amplitude A (see Equation 2.18 and Equation 2.20). The cumulative distribution for the negative exponentially distributed random variable is found by integrating Equation 2.20, with $<I> = 2$ in this case so that the Cartesian components have unit variance, which gives

$$F_I\left(I\right) = 1 - \exp\left(-I / 2\right) \tag{14.6}$$

Solving this for I gives the required transformation, via Equation 14.4,

$$A = \sqrt{I} = \sqrt{1 - 2\log\left(1 - x_2\right)} \tag{14.7}$$

y_1 is then obtained from

$$y_1 = A\cos\left(\phi\right) = \cos\left(2\pi x_1\right)\sqrt{1 - 2\log\left(1 - x_2\right)} \tag{14.8}$$

Term y_2 is obtained similarly, using $\sin(\phi)$ instead of $\cos(\phi)$. Terms y_1 and y_2 are zero-mean uncorrelated Gaussian random numbers with unit variance. Other variances can be obtained by using suitable multiplying factors, and numbers with nonzero mean can be obtained by adding a constant. The complex number $z = y_1 + iy_2$ is a circular complex Gaussian variable.

The methods discussed above produce single random numbers. If they are applied multiple times, a sequence of uncorrelated numbers is produced. In general, however, simulation of a random process requires an array of random numbers which have some degree of correlation with one another.

14.4 The Telegraph Wave

A random process that can be easily simulated is the random telegraph wave (or telegraph signal) discussed in Chapter 8; this is also sometimes referred to as a dichotomous random process since it switches between two values randomly. The simplest such process is one in which the transitions between states occur with equal probability at all times and are therefore mutually independent [2]. A simulated telegraph wave is shown in Figure 8.5. Each point in the array of numbers plotted represents the value of the telegraph wave during a discrete time interval δt. The probability that there will be a change in the state of the telegraph wave during this interval is taken to be $\gamma\delta t$. The simulation operates as follows: for each time step a random number is generated which is uniformly distributed between 0 and 1, only if this number is less than $\gamma\delta t$ is the state of the telegraph wave changed from its previous value. For this to be an accurate representation of a continuous-time telegraph wave, the probability $\gamma\delta t$ must be sufficiently small that the chances of two, or more, changes of state occurring during one time interval are negligible. The probability of m events occurring is governed by the Poisson distribution of Equation 13.44

$$P\left(m,\gamma\delta t\right) = \frac{\left(\gamma\delta t\right)^{m}}{m!}\exp\left(-\gamma\delta t\right) \tag{14.9}$$

Thus, the probability of more than one change of state is $1-(1 + \gamma\delta t)\exp(-\gamma\delta t)$, which for small $\gamma\delta t$ is $\approx (\gamma\delta t)^2$. For the data in Figure 8.5 a value of $\gamma\delta t = 0.01$ was used, so there is only one chance in 10^4 per interval of a multiple event being missed. Of course, the degree of accuracy required from a simulation depends on its actual application and needs to be decided accordingly.

14.5 The Gaussian Random Process

A class of signals that can be readily simulated are the Gaussian random processes discussed in Section 2.5. Simulation is greatly facilitated by the special property of Gaussians variables — the fact that a linear combination of Gaussian variables is also a Gaussian variable (Section 2.3). It was shown in Chapter 2 that two correlated Gaussian variables could be produced by linear combination of a pair of independent Gaussian variables (Equation 2.4). This can be generalized to a vector of correlated Gaussian variables u being produced by multiplying a vector v of independent Gaussian variables by a matrix $\mathbf{A}$.

$$\bar{u} = \mathbf{A}\bar{v} \tag{14.10}$$

The correlations between the components of $\bar{u}$ are expressed by a correlation matrix $\boldsymbol{\rho}$ given by

$$\rho = \tilde{A}^* A \tag{14.11}$$

where $\tilde{A}^*$ is the complex conjugate of the transpose of A. The correlation matrix for Equation 2.4 can be expressed as

$$\rho = \left\{ \begin{array}{cc} 1 & \sin(2\theta) \\ \sin(2\theta) & 1 \end{array} \right\} \tag{14.12}$$

when the u and v elements are of unit variance.

A very useful method for simulating stationary Gaussian random processes is based on the discrete Fourier transform (DFT). The elements of the transforming matrix A are expressed in terms of the Fourier transform of the required correlation function, and $\bar{u}$ is calculated by a numerically efficient DFT (i.e., a fast Fourier transform or FFT). This method is sometimes referred to as filtering of Gaussian noise, because it is equivalent to starting with white (i.e., spectrally uniform) noise and imposing the required correlation via a filter (a multiplication in the frequency domain).

The DFT is analogous to the ordinary Fourier transform, but operates on discrete arrays of numbers rather than continuous functions. All real data is discrete, and DFTs find many applications in numerical analysis.

Consider a one-dimensional array of N numbers $\{f_n\}$. The DFT of this is another array of length N, $\{F_m\}$, defined by

$$F_m = \frac{1}{\sqrt{N}} \sum_{n=0}^{N-1} f_n \exp\left(i \frac{2\pi mn}{N} \right) \tag{14.13}$$

The inverse transformation is

$$f_n = \frac{1}{\sqrt{N}} \sum_{m=0}^{N-1} F_m \exp\left(-i \frac{2\pi mn}{N} \right) \tag{14.14}$$

A detailed discussion of the properties of the DFT is beyond the scope of this book. There are many references that the interested reader can consult ([3], for example); those features that are relevant to the problems considered here will be discussed as they arise. The similarity to the ordinary Fourier transform (Equation 14.15 below) is evident. A continuous function $f(x)$ can be *sampled* at intervals Δx to produce a sequence $\{f_n\}$ from which a transformed sequence $\{F_m\}$ can be derived via Equation 14.13. Under ideal conditions this will be a good approximation to the Fourier transform of f, sampled at multiples of $\Delta k = 2\pi/(N\Delta x)$. There is a multiplying factor when going between the DFT and the continuous Fourier transform which depends on the definitions used (a number of definitions are in common usage). When using Equation 14.13 and Equation 14.14 to define the DFT and

$$F(k) = \int_{-\infty}^{\infty} f(x)\exp(ikx)\,dx$$

$$f(x) = \frac{1}{2\pi} \int_{-\infty}^{\infty} F(k)\exp(-ikx)\,dk$$

(14.15)

for the continuous transform, the multiplying factor is $\Delta x \sqrt{N}$, that is, under ideal conditions the samples of F are given by the numbers F_m multiplied by $\Delta x \sqrt{N}$. However, significant information about the function f (and thus F also) will obviously be lost in this process if the function is nonzero outside the range of x values used in the sampling, or if it varies too greatly within any Δx interval.

The random array representing a realization of a Gaussian process is produced by the inverse DFT of the product of a set of weights, derived from the required autocorrelation function, and an array of uncorrelated complex Gaussian numbers of zero mean and unit variance

$$z_n = \frac{1}{\sqrt{N}} \sum_{m=0}^{N-1} a_m w_m \exp\left(-i\frac{2\pi mn}{N}\right)$$

(14.16)

The numbers a_m can be generated using the method of Section 14.3. Since z_n is a sum of Gaussians, it is itself a complex Gaussian number. In fact, z_n is a sample of a circular Gaussian process, since its real and imaginary parts are uncorrelated and of equal variance (this can be easily shown using Equation 14.16).

Since the random numbers a_m are uncorrelated, that is, $\langle a_m a_q^* \rangle = \delta_{mq}$, the correlations of z are given by

$$\langle z_n z_p^* \rangle = \frac{1}{N} \sum_{m=0}^{N-1} w_m^2 \exp\left(-i\frac{2\pi m(n-p)}{N}\right)$$

(14.17)

By comparing this with Equation 14.14 it can be seen that the required weights $\{w_m\}$ are related to the autocorrelation through Equation 14.13

$$w_m^2 = \sum_{r=0}^{N-1} \langle z_n z_{n-r}^* \rangle \exp\left(i\frac{2\pi mr}{N}\right)$$

(14.18)

Thus, one can start with the required autocorrelation function and calculate its DFT, and the squares of the required weights are the DFT elements multiplied by $\sqrt{N}$. Alternatively, since the continuous Fourier transform of the autocorrelation function is the power spectral density of z, $S_z(k)$ (see Section 1.4), one can express the weights as $w_m^2 = S_z(m\Delta k)/\Delta x$, making use of the multiplying factor discussed previously to go from the continuous Fourier transform to the DFT.

Note that in Equation 14.17 the difference $r = n - p$ can take on negative values, whereas in the original definition (Equation 14.14), positive values were assumed. Since the complex exponential is periodic, a negative value of r gives the same result as a value of $N - r$. This "wrap-around" is a fundamental property of the DFT, and appears in many guises. In the present case it is connected to a property of the realizations of the random process generated via the DFT: that of circular symmetry. The final point in the array of random values differs from the first by a factor of $\exp(2\pi im/N)$ in each term of the summation in Equation 14.16. However, exactly the same difference applies to each consecutive pair in the array; so the last and first values behave like any other consecutive pair and the array can be thought of as forming a continuous loop, its beginning following on from its end.

If a real Gaussian process is required, the real or imaginary part of z can be used. Note that the variance of the resulting process will be 0.5 (assuming that one starts with the normalized autocorrelation function). An example is given in Figure 14.1, in which a short segment of random process is generated. This starts with a Gaussian function for the normalized autocorrelation function, shown in the upper plot. Note that the zero delay point is the first point in the array, and that the negative delays start from the end of the array, as discussed above. The number of points in the array is 512, which is a power of two; this gives fastest FFT operation. In this example, the largest positive delay is at position 257 in the array, which is immediately followed by the largest negative delay at position 258. The real part of the generated realization is plotted below the autocorrelation function. Note the way in which the end of the time series joins up with the beginning. If this circular correlation is not desired, then at least one correlation length must be dropped from the end or beginning of the time series. The realizations of one-dimensional Gaussian processes in Figure 1.1, Figure 1.3, and Figure 2.1 were produced in this way. The trace in Figure 1.5 started with a complex Gaussian process: the real part was multiplied by a cosine wave, the imaginary part by a sine wave, and the two summed. The lines indicating the envelope were produced from the absolute value of the complex time series. Figure 2.6 plots the argument of a complex time series.

The extension to two-dimensional arrays is straightforward: two DFTs are applied, one to rows and one to columns; this lends itself to Cartesian coordinate systems. An example is shown in Figure 14.2, where the real part of a two-dimensional, complex realization is plotted. This example uses a square array of 256 by 256 points, but rectangular arrays can be produced in the same way and can be useful when simulating a phase screen which

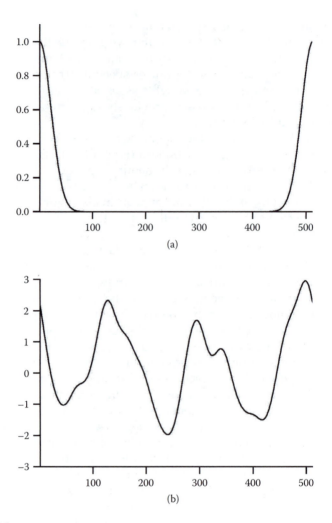

FIGURE 14.1
Upper plot is a Gaussian autocorrelation function with origin at the start of the array and negative delays starting at the end, in order to conform to the DFT convention; lower plot is a realization of a Gaussian process having this autocorrelation function, note continuity in lower plot between the final and first points of the array.

is moving perpendicular to the propagation direction: the long axis can lie along the direction of motion, which makes the most efficient use of the generated phase screen when the illuminated beam has the same dimensions in the x and y directions, since the short axis only needs to be long enough to accommodate the beam. Figure 14.2 was produced for a Gaussian autocorrelation function following the same procedure used for the one-dimensional example of Figure 14.1, but applied in both dimensions. The radius of the Gaussian autocorrelation function was 2 cm, with the grid spacing taken to be 1 mm.

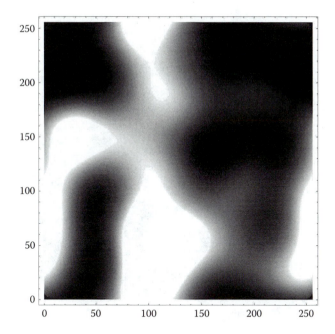

FIGURE 14.2
Two-dimensional Gaussian process with a Gaussian autocorrelation function.

A further two-dimensional example is given in Figure 14.3. This started from the spectrum rather than the autocorrelation function. The power spectral density was that of Equation 9.30, the Von-Karman spectrum, substituted in Equation 9.6 to give

$$S_\phi(\mathbf{k}) = \frac{8.187\, z k^2 C_n^2}{\left(|\mathbf{k}|^2 + \left(\dfrac{2\pi}{l_o}\right)^2\right)^{\frac{11}{6}}} \exp\left(-|\mathbf{k}|^2 \left(\frac{l_i}{5.92}\right)^2\right) \qquad (14.19)$$

The difference between the single-scale nature of the Gaussian and the fractal nature of the Von Karman can be seen clearly in Figure 14.3: there is structure on a wider range of scales.

When the inner scale, l_i, and the outer scale, l_o, are sufficiently well separated, the structure function of the variable ϕ has a region with a 5/3 power law, which is characteristic of Kolmogorov turbulence (Equation 9.31). Note that it is not possible to simulate a pure power-law process in this way because the spectrum needs to have well-defined high and low frequency cutoffs in order for it to be sampled correctly for the DFT. The data plotted in Figure 14.3 consists of an array of 512 by 512 points and was produced as follows: taking the separation between points to be 1 mm gives a spatial

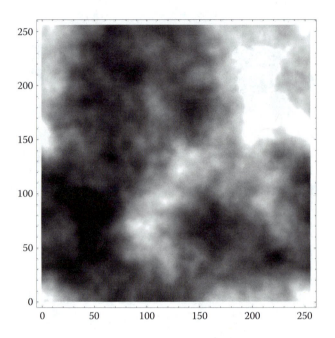

FIGURE 14.3
Two-dimensional Gaussian process with a von Karman spectrum.

frequency step size of $\Delta k = 2\pi/(N\Delta x) = 12.27$ m^{-1}. The inner scale was set to 5.3 mm and the outer to 63 cm, the value of C_n^2 was 10^{-14} m$^{-2/3}$, and the wavelength was 600 nm. In terms of array indices p and q, $|\mathbf{k}|^2$ in Equation 14.19 is replaced by $\Delta k^2((p - 128)^2 + (q - 128)^2)$. This gives an array with the origin at the center; the array is reshaped to move this point to the first position in both the column and row directions. That is, both the rows and columns are reordered in the same way as in the one-dimensional example discussed previously, with the negative frequencies counting back from the end of the row/column. This reshaping is shown in Figure 14.4 using the example of a two-dimensional Gaussian function. The power spectral density values are then divided by Δx^2, one factor of Δx for each array dimension, to get the squares of the required weights. These are then square rooted and multiplied, point by point, by an array of uncorrelated complex Gaussian random numbers of zero mean and variance of two, which gives real and imaginary parts of unit variance. The inverse DFT is applied and the real part taken to give the result of Figure 14.3. Note that the imaginary part gives another independent realization of the process.

In Figure 14.3 the outer scale value used is somewhat shorter than the length of the array. It is not possible, using this method, to have an outer scale larger than the array length: if such a value is used in Equation 14.19 an outer scale with a magnitude on the order of the array length is imposed instead by the circular correlation, which can be seen in the continuity across the boundaries of the array in Figure 14.2 and Figure 14.3. Methods for

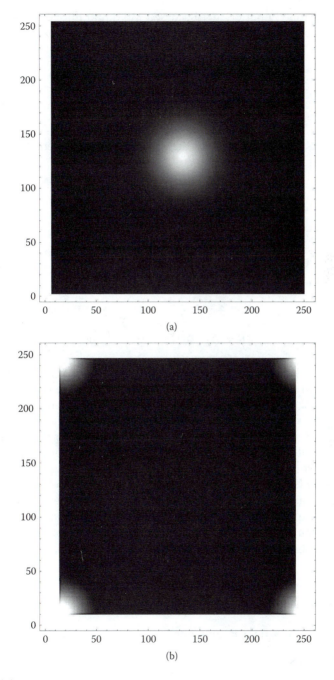

(a)

(b)

FIGURE 14.4
Reshaping of a two-dimensional Gaussian function (with the origin at the center of the plot) to conform to the DFT convention, where the origin is at the lower left corner.

overcoming this limitation are discussed in [4] and [5]. For some classes of Gaussian process, other methods can be used which do not impose an outer scale in this way; an example is the Brownian fractal process of Chapter 7, the structure function of which is

$$\left\langle [\phi(t) - \phi(t+\tau)]^2 \right\rangle = 2\alpha\,|\tau| \tag{14.20}$$

This process can be written as the solution of a differential equation (a Langevin equation)

$$\frac{d\phi}{dt} = \sqrt{2\alpha}\,\varepsilon(t) \tag{14.21}$$

Here ε is a zero-mean, unit variance, delta-correlated Gaussian variable, so that $\langle\varepsilon(t)\varepsilon(t+\tau)\rangle = \delta(\tau)$, $\delta(\tau)$ being the Dirac delta function. This can be simulated by putting it into integral form

$$\phi(t+\Delta t) = \phi(t) + \sqrt{2\alpha}\int_{t}^{t+\Delta t}\varepsilon(t')\,dt' \tag{14.22}$$

For sufficiently small Δt the integral can be replaced by a Gaussian variable of variance Δt, and the simulation proceeds by incrementing the variable ϕ at every time step by a random amount, the random increment being Gaussian distributed with variance $2\alpha\Delta t$. The fractal nature of this model means that numerical simulations will never be exact because there is no inner scale. The time step needs to be chosen such that the resulting loss of fine scale structure is of negligible impact: its value will depend on the application. As with all such simulations, it is a good idea to run the calculation twice, with the value of Δt halved on the second occasion, to check that a consistent result is obtained. An example of a simulated Brownian fractal is given in Figure 14.5, the product $\alpha\Delta t$ being 0.02.

14.6 Non-Gaussian Processes

Some non-Gaussian processes can be simulated in a straightforward manner because they arise from simple models; an example is the telegraph wave, discussed in Section 14.4. Others are themselves based on underlying Gaussian random processes, and can be fully specified by the correlation properties of the underlying Gaussian; examples are the Rice random process

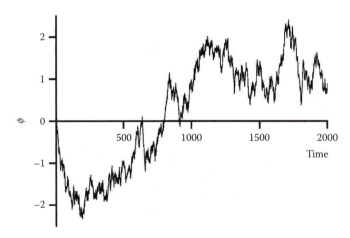

FIGURE 14.5
A Brownian fractal process.

and the lognormal random process discussed in Chapter 3. These can be simulated by starting with a jointly Gaussian random process and applying a specified transformation. Simulation of arbitrary non-Gaussian processes is, however, more problematic.

In fact, the jointly Gaussian random process is a very special case, as it is completely defined by its correlation function. This is not, in general, true for non-Gaussian processes. For these, a given marginal density and correlation function can be possessed by an arbitrary number of different processes, the differences appearing in the higher-order statistics. It is more difficult to use non-Gaussian models because of this. For example, if in a phase screen calculation the random phase is specified as having a given correlation function and non-Gaussian marginal density, it will be incompletely described, because many scattering results will depend on the unspecified higher-order statistics. This differs from the jointly Gaussian case when specification of a correlation function for the phase is sufficient to determine all higher-order statistics; in the non-Gaussian case the higher-order statistics of any model would have to be decided by consideration of the detailed physical processes that give rise to the random behavior.

In some situations, experimental data or heuristic arguments may suggest that a certain marginal density is to be expected for a random variable, without fully constraining the higher-order statistics. Examples are the *K*-distribution used to model microwave returns from the sea surface (see Chapter 12) and the various distributions proposed to describe intensity fluctuations in propagation through atmospheric turbulence (see Chapter 9). Techniques have been developed to simulate such non-Gaussian random processes based on just marginal statistics and correlation functions [6–9]. Some simulated quantities may depend on these low-order properties only, the higher-order statistics being irrelevant when these are calculated. Other quantities will be, to a greater or lesser extent, dependent on higher-order statistics and will thus vary depending on the particular simulation technique used. This is somewhat unsatisfactory, since the higher-order statistics will be determined by the implementation of a particular simulation technique rather than on the basis of any experimental data or theoretical concept.

The remainder of this section will illustrate the points discussed above by considering one popular simulation technique: the memory-less nonlinear transformation of a Gaussian process (MNLT), also called the zero-memory nonlinearity (ZMNL) technique [7–9]. The approach is to start with a jointly Gaussian random process, generated as in Section 14.5, and to apply a nonlinear transformation to each generated number to give a series of values having the desired marginal density. The problem then becomes that of finding the correlation function for the initial Gaussian random process that leads to the desired correlation function for the transformed process.

Starting with a zero-mean, unit-variance, real Gaussian variable x, Equation 14.3 gives the required transformation as

$$y = f(x) = F_y^{-1}\left(\text{erf}\left(x/\sqrt{2}\right)/2 + 1/2\right) \tag{14.23}$$

where F_y^{-1} is the inverse of the cumulative distribution function of the desired variable y, and the error function on the right-hand side arises from the cumulative distribution of x, which is an integral over the Gaussian pdf. The correlation function of y can be expressed in terms of the joint pdf of the Gaussian variable

$$\left\langle y^2 \right\rangle \rho_y(r) = \iint f(x)f(x')P(x,x')\,dx\,dx' \tag{14.24}$$

the values x and x' being separated by the interval r, which could represent a spatial variable or a delay time. The term $P(x,x')$, given by Equation 2.12, can be expressed as a power series in ρ_x, the correlation function of x, [8] leading to

$$\left\langle y^2 \right\rangle \rho_y(r) = \frac{1}{2\pi} \iint f(x)f(x')\exp\left(-\frac{x^2 + x'^2}{2}\right)$$
$$\sum_{n=0}^{\infty} \frac{H_n\left(\dfrac{x}{\sqrt{2}}\right)H_n\left(\dfrac{x'}{\sqrt{2}}\right)}{2^n n!}\left[\rho_x(r)\right]^n dx\,dx' \tag{14.25}$$

where the H_n are Hermite polynomials. Integrating term by term gives

$$\left\langle y^2 \right\rangle \rho_y (r) = \frac{1}{2\pi} \sum_{n=0}^{\infty} \frac{\left[\rho_x (r) \right]^n}{2^n n!} \left[\int H_n \left(\frac{x}{\sqrt{2}} \right) f(x) \exp \left(-\frac{x^2}{2} \right) dx \right]^2 \quad (14.26)$$

For a given $f(x)$ the integral on the right-hand side of Equation 14.26 can be evaluated, numerically if necessary, to obtain the coefficients of the power series. In most cases of practical interest, only the first few terms of the series are required [9]. Inverse tabulation can then be applied to find $\rho_x(r)$ as a function of $\rho_y(r)$.

As an example, consider the gamma distributed phase screen discussed in Chapter 8. The MNLT method can be used to generate one-dimensional phase screens with gamma-distributed marginal density and Gaussian correlation function. By averaging over many realizations, the second moment of intensity can be calculated as a function of distance from the screen. However, this quantity depends on the four-point joint pdf of phases at different positions on the screen [10], so it is a function of higher-order statistical properties; this is illustrated below.

For integer values of the parameter 2α (see Equation 3.12), two different methods can be used to generate the phase ϕ. The MNLT method is one, and the other is to start with the appropriate number of Gaussian variables and sum their squares; that is, starting with independent identically distributed Gaussian variables X_n, of unit variance and correlation function $g(r)$. Using this second approach, the phase is given by Equation 3.11

$$\phi = \frac{1}{2\alpha} \sum_{n=1}^{2\alpha} X_n^2 \quad (14.27)$$

and the correlation function is given by Equation 3.19. Thus a Gaussian correlation function for ϕ results if the X_ns also have identical Gaussian correlations; the width is reduced by $\sqrt{2}$ because of the power of two in Equation 3.19. The fact that the mean of ϕ is nonzero is unimportant because it just gives a constant phase shift which doesn't change the intensity fluctuations produced by the phase screen. This second method will give results conforming to the model of Chapter 8. Note, however that this approach is not applicable to noninteger values of 2α, whereas the MNLT technique is.

Consider the case of $\alpha = 1$, which gives a negative exponential marginal density. For this case, ϕ can be generated by the MNLT method or as the sum of the squares of two independent Gaussian random variables. Examples of realizations, with Gaussian correlation function, using the different techniques are plotted in Figure 14.6. Also shown is the result for a Gaussian variable with the same correlation function. Close inspection does reveal a subtle different in the Gaussian variable; there is a symmetry between positive and negative values that is lacking in the other plots. However, any differences between the other two are not so evident.

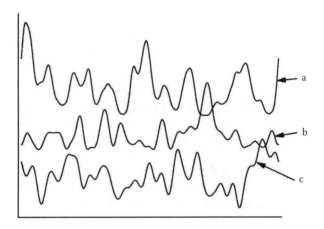

FIGURE 14.6
a) Gamma process generated by the MNLT method; b) Gamma process generated by summing the squares of two Gaussians; c) Gaussian process with the same correlation function.

It is instructive to examine the two-point joint pdfs in the two cases. For the sum of squares method the joint pdf is given by Equation 3.18 with $\alpha = 1$.

$$P_1\left(y_1, y_2\right) = \frac{1}{2 - \rho_y} \exp\left(-\frac{y_1 + y_2}{2 - \rho_y}\right) I_0\left(\frac{2\sqrt{\left(\rho_y - 1\right)y_1 y_2}}{2 - \rho_y}\right) \qquad (14.28)$$

In the MNLT method the required nonlinear transform is

$$f\left(x\right) = \ln 2 - \ln\left(1 - \mathrm{erf}\left(x/\sqrt{2}\right)\right) \qquad (14.29)$$

The joint pdf can be found from Equation 2.7, noting that the Jacobian for the transformation of the joint density can be written in terms of the marginal densities as

$$|J| = \frac{P_x\left(x_1\right)P_x\left(x_2\right)}{P_y\left(y_1\right)P_y\left(y_2\right)} \qquad (14.30)$$

Thus the joint pdf is

$$P_2\left(y_1, y_2\right) = \frac{\exp\left(-y_1 - y_2\right)}{\sqrt{1 - \rho_x^2}} \exp\left(\frac{2x_1 x_2 \rho_x - \rho_x^2\left(x_1^2 + x_2^2\right)}{2\left(1 - \rho_x^2\right)}\right) \qquad (14.31)$$

with $x_{1,2} = \sqrt{2}\,\mathrm{InvErf}\left(1 - 2\exp\left(-y_{1,2}\right)\right)$ and with the correlation function ρ_x related to ρ_y by the required transformation. In this case the transformation can be approximated to a high degree of accuracy by the four-term power series

$$\rho_x = -1.56 + 1.96\rho_y - .46\left(\rho_y\right)^2 + 0.06\left(\rho_y\right)^3 \qquad (14.32)$$

These joint pdfs are functions of three variables: y_1, y_2, and ρ_y. Some values are plotted in Figure 14.7. Here the joint densities for fixed values of y_1, y_2 are plotted as a function of the correlation coefficient ρ_y. It can be seen that the two functions are similar, but not identical. The significance of these differences in any application will depend on what is being calculated.

Figure 14.8 shows simulation results for the second moment of intensity, in a 1-D phase screen scattering geometry, that is, a corrugated phase screen (see Section 14.7 below for the relevant methods). This quantity is completely determined by the four-point joint pdf (see Reference [10]). Results for a Gaussian phase screen with the same correlation are also given. Here, the variance of the phase shift was 400 and the correlation length ($1/e$) was 240 spatial steps. It can be seen that there is a significant difference between the results using the two different methods of generating this negative exponential process; these arise from the differences in the joint pdfs. The maximum value of the second moment is greater when the MNLT method is used. The shape of the curve of the second moment is, in both cases, clearly different from the Gaussian result. So, qualitative features of the negative exponential phase screen are being captured by both approaches, but higher-order statistics are important in determining the exact values of the second moment.

14.7 Simulation of Wave Propagation

The simulation of wave propagation is another area in which the FFT is of great utility. It allows one to evaluate the Huyghens-Fresnel integral of Equation 9.18 more efficiently than a direct numerical integration. Equation 9.18 is the solution to the paraxial wave equation when the field is given over an infinite halfplane at $z = z_0$. This equation is

$$\frac{\partial^2 E}{\partial x^2} + \frac{\partial^2 E}{\partial y^2} - 2ik\frac{\partial E}{\partial z} = 0 \qquad (14.33)$$

Here, z is the propagation direction and k is the wave vector ($2\pi/\lambda$, λ being the wavelength) of the monochromatic electromagnetic wave with slowly varying electric field E. By taking a 2-D Fourier transform with respect to

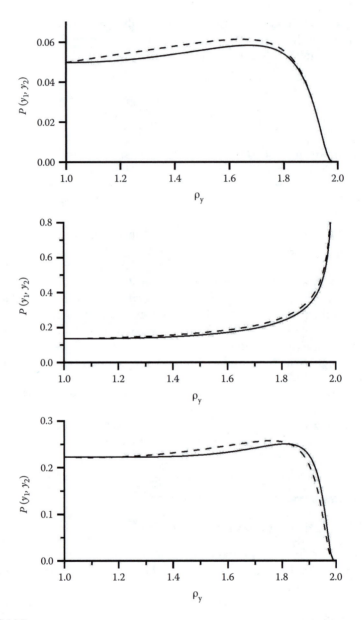

FIGURE 14.7
Comparison of the joint pdfs for two different Gamma processes (dashed line is MNLT version); upper trace: $y_1 = 1$, $y_2 = 2$, middle trace: $y_1 = 1$, $y_2 = 1$, lower trace: $y_1 = 1$, $y_2 = 0.5$.

the x and y coordinates, the following expression for the transform of the field at z_1 produced by a field at z_0 is obtained

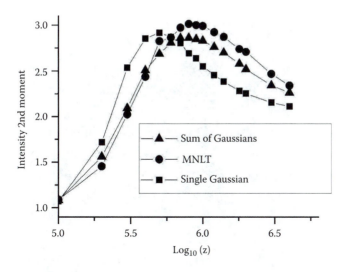

FIGURE 14.8
Calculations of second moment of intensity as a function of distance for scattering by corrugated negative exponential phase screens.

$$e(\mathbf{k}, z_1) = e(\mathbf{k}, z_0) \exp\left[\frac{i|\mathbf{k}|^2 (z_1 - z_0)}{2k}\right] \qquad (14.34)$$

Here, the vector $\mathbf{k}$ is the transform variable (the spatial frequency). This equation can be evaluated via the DFT. Note that for a corrugated screen the y differential in Equation 14.33 can be set to zero, and a one-dimensional DFT used; this method was used to produce the data in Figure 14.8. Since the terms on both sides of Equation 14.34 have undergone a Fourier transform, multiplying factors are not required when converting to the DFT, as they appear on both sides of Equation 14.34 and thus cancel. Therefore, the equivalent equation for the DFTs of the fields is just a discrete version of Equation 14.34, and the field at z_1 is obtained via the inverse transform. A further advantage of using the DFT, in addition to computational speed, is that optical power is automatically conserved, that is, the modulus squared of the field summed over the array is conserved (this can be readily shown using Equation 14.34 as well as Equation 14.13, Equation 14.14, and Equation 14.15).

The simulation is implemented by representing the field by a square grid of $n \times n$ elements, with a value of the field assigned to each array element. Again, the origin is placed at the first point in the array, as in the lower plot of Figure 14.4. The application of two DFTs, one for rows and one for columns, transforms this $n \times n$ array in the space domain to an $n \times n$ array in the Fourier domain, which is then multiplied by another array of values from the exponential term in Equation 14.34. This term is sampled by first

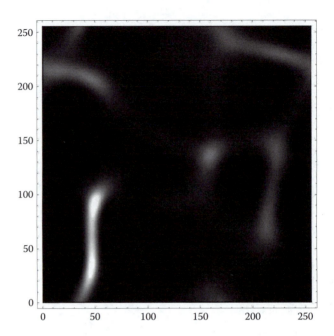

FIGURE 14.9
Intensity pattern produced by scattering by a Gaussian-correlated phase screen.

replacing $|\mathbf{k}|^2$ by $\Delta k^2((p - n/2)^2 + (q - n/2)^2)$, p and q being the row and column positions, and then shifting the origin to the first point in the array, as described previously. Finally, the inverse DFT gives the field at z_1.

An example is shown in Figure 14.9. This shows the Fresnel regime of a plane wave that has been scattered by the Gaussian-correlated phase screen of Figure 14.2. The rms phase shift was 10, the wavelength 600 nm, and the propagation distance 1km. Each point in the two-dimensional phase screen array was multiplied by the square root of minus one and exponentiated. The procedure described above was then used to calculate the complex field after propagation to the distance of 1 km, the modulus-squared of which gives the intensity pattern shown in Figure 14.9. This propagation distance corresponds to the focusing regime (see Chapter 6; the intensity pattern here is similar to those in Figure 6.2a and Figure 6.2b).

Clearly, for the simulation to be accurate, both the initial field and the exponential term in Equation 14.34 need to adequately sampled. Since the argument of the exponential term is imaginary, the term oscillates as a function of $|\mathbf{k}|^2$, and as the propagation distance $z_1 - z_0$ is increased the oscillations get more rapid. So, if Δk is not reduced, this term becomes less well sampled as the propagation distance is increased because there is more variation occurring in between the points at which it is sampled. The consequence of this undersampling in the spatial-frequency domain is actually very simple: in the spatial domain the propagated wave goes outside of the region on which the square grid is defined; however, because of the periodic nature of the DFT, light which propagates outside of the sampled region wraps around and reappears across the opposite boundary. Another way of thinking of this is to realize that the periodic nature of the DFT means that this method is actually equivalent to simulating the propagation of an infinite array of identical square regions. When the propagation is extended to sufficient distance, light starts to spill over into neighboring squares. This is shown in Figure 14.10, where the Gaussian profiled beam in Figure 14.10a is allowed to propagate until it diffracts outside of the sampled region. The light which overspills the boundary produces fringes when it interferes with light that has remained within the original square region; this is seen in Figure 14.10b. It is instructive to apply this concept to a plane-wave simulation such as that in Figure 14.9. It is, in fact, only the circular correlation of the phase screen used in the simulation (i.e., that of Figure 14.2) that prevents diffraction of the wave by discontinuities at the edge of the sampled region producing immediate overspill and spurious interference fringes. However, at sufficiently large propagation distances, spurious interference effects will still occur when light from one inhomogeneity is diffracted by an amount on the order of the width of the sampled region. That is, the results will be affected by the periodicity inherent in the simulation and will differ from those that would be obtained from an infinite, nonperiodic phase screen. These effects can be avoided by using a larger grid, and thus a smaller Δk.

14.8 Multiple Phase Screens

As mentioned in Chapter 9, multiple phase screens can be used to model propagation through extended media. This technique has often been applied to the simulation of propagation through the turbulent atmosphere [4,11,12,13]. The simplest approach is to represent the extended medium by a set of equally spaced screens; although for nonuniform turbulence, particularly when the turbulence strength is varying along the propagation path, a more effective approach may be to use unequal spacings. An example of the intensity pattern produced by propagation through a medium with a von Karman spectrum is given in Figure 14.11. The parameters are the same as the section of medium represented by the single phase screen in Figure 14.3, but 20 equally spaced screens were used, each with one-twentieth of the phase variance. These were generated in pairs using the real and imaginary parts of a complex Gaussian realization. The first screen was placed at the start of the medium, and the inter-screen distance was one-twentieth of the total path: 50 meters. In this plane-wave scenario, any effects arising from wraparound, as in Figure 14.10, are less obvious, so a check was carried out by propagating a Gaussian beam through the same set of screens and checking that there was no significant overspill.

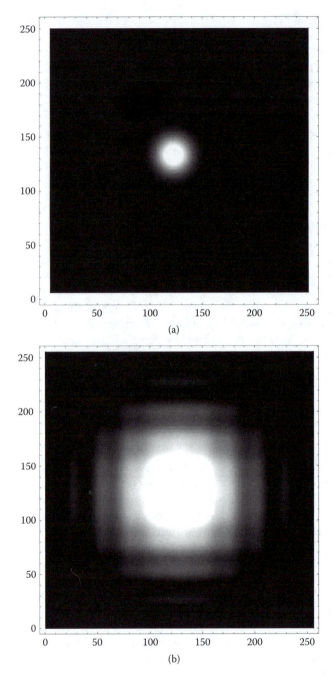

FIGURE 14.10
Effect of wraparound on propagation via DFT method.

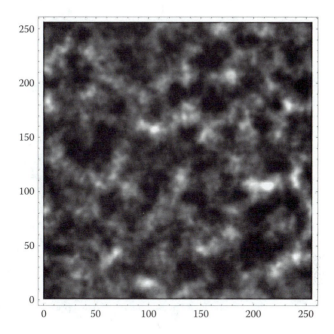

FIGURE 14.11
Intensity pattern produced by propagation through Kolmogorov turbulence.

One obvious question that arises is: how many screens are required to adequately represent the medium? Figure 14.12 shows the results of an empirical investigation of this, for the scenario of Figure 14.11. The scintillation index is plotted as a function of the number of screens used. This was calculated using 100 realizations for each number of screens. The scintillation index is small, indicating weak scattering. As expected, using a very small number of screens results in an overestimate of the index. The value drops as the number of screens is increased, reaching a reasonably constant level when 20 screens are used. The solid line is the Rytov result of Equation 9.32 which predicts a value somewhat greater than that produced in the simulation.

Clearly, time dependence may be introduced by moving the phase screens and recalculating the propagation after each time step. Uniform and equal transverse motion of all screens would give a result conforming to Taylor's frozen flow hypothesis, but other forms of motion, including motion along the beam propagation direction, can also be simulated in this way.

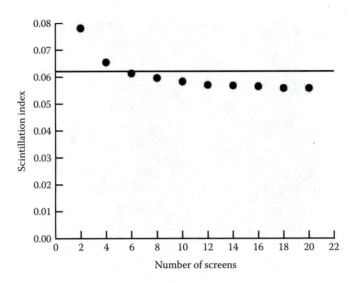

FIGURE 14.12

Scintillation index as a function of number of phase screens used to represent extended medium.

References

1. Knuth, D.E. *The Art of Computer Programming: Volume 2 Seminumerical Algorithms.* 3rd ed. Boston: AddisonWesley, 1998.
2. Goldstein, S. On diffusion by discontinuous movements and telegraph equations. *Quart. J. Mech. Appl. Math IV* (1951): 129–56.
3. Weaver, H.J. *Theory of Discrete and Continuous Fourier Analysis.* New York: Wiley, 1989.
4. Frehlich, R. Simulation of Laser Propagation in a Turbulent Atmosphere. *Appl. Opt.* 39 (2000): 393–97.
5. Lane, R.G., A. Glindemann, and J.C. Dainty. Simulation of a Kolmogorov Phase Screen. *Waves in Random Media* 2 (1992): 209–24.
6. Johnson, G.E. Constructions of Particular Random Processes. *Proc. IEEE* 82, no. 2 (1994): 270–85.
7. Liu, B., and D.C. Munson. Generation of a Random Sequence Having a Jointly Specified Marginal Distribution and Autocovariance. *IEEE Trans. Acoust. Speech and Sig. Process.* ASSP-30 (1982): 973–83.
8. Sondhi, M.M. Random Processes with Specified Spectral Density and First-Order Probability Density. *Bell Sys. Tech. J.* 62, no. 3 (1982): 679–701.
9. Tough, R.J.A., and K.D. Ward. The Correlation Properties of Gamma and Other Non-Gaussian Processes Generated by Memoryless Nonlinear Transformation. *J. Phys.* D 32 (1999): 3075–84.
10. Jakeman, E. Scattering by Gamma-Distributed Phase Screens. *Waves in Random Media* 2 (1991): 153–67.
11. Knepp, D.L. Multiple Phase-Screen Calculation of the Temporal Behaviour of Stochastic Waves. *Proc. IEEE* 71 (1983): 722–37.

12. Martin, J.M., and S.M. Flatté. Intensity Images and Statistics from Numerical Simulation of Wave Propagation in 3-D Random Media. *Appl. Optics* 27 (1988): 2111–26.
13. Flatté, S.M., and J.S. Gerber. Irradiance-Variance Behaviour by Numerical Simulation for Plane-Wave and Spherical-Wave Optical Propagation through Strong Turbulence. *J. Opt. Soc. Am.* A 17 (2000): 1092–97.

Index

C